Instant Notes

Bioinformatics

The INSTANT NOTES series

Series editor
B.D. Hames
School of Biochemistry and Molecular Biology, University of Leeds, Leeds, UK

Animal Biology
Biochemistry 2nd edition
Chemistry for Biologists
Developmental Biology
Ecology 2nd edition
Genetics
Immunology
Microbiology
Molecular Biology 2nd edition
Neuroscience
Plant Biology
Psychology
Bioinformatics

The INSTANT NOTES Chemistry series
Consulting editor: Howard Stanbury

Analytical Chemistry
Inorganic Chemistry
Medicinal Chemistry
Organic Chemistry
Physical Chemistry

Bioinformatics

David R. Westhead, J. Howard Parish

School of Biochemistry and Molecular Biology,
University of Leeds, UK

and

Richard M. Twyman

Department of Biology,
University of York, UK

ISBN 1 85996 272 6

BIOS Scientific Publishers Ltd
9 Newtec Place, Magdalen Road, Oxford OX4 1RE, UK
Tel. +44 (0)1865 726286. Fax +44 (0)1865 246823
World Wide Web home page: http://www.bios.co.uk/

Distributed exclusively in the United States, its dependent territories, Canada, Mexico, Central and South America, and the Caribbean by Springer-Verlag New York Inc, 175 Fifth Avenue, New York, USA, by arrangement with BIOS Scientific Publishers, Ltd, 9 Newtec Place, Magdalen Road, Oxford OX4 1RE, UK

Production Editor: Andrew Watts
Typeset by Phoenix Photosetting, Chatham, Kent, UK
Printed by Biddles Ltd, Guildford, UK, www.biddles.co.uk

Cover illustration from Jackson R.M., Chapter 13, 'Predicting the structure of protein-biomolecular interaction' in Orengo, Jones and Thornton (eds), *Bioinformatics: genes, proteins and computers*, ©BIOS Scientific Publishers 2002.

CONTENTS

ABBREVIATIONS

2D-PAGE	two-dimensional polyacrylamide gel electrophoresis	FRET	fluorescent resonance energy transfer
AC	approximate correlation	FTP	file transfer protocol
ACeDB	A C. elegans DataBase	GASP	Gene Annotation aSsessment Project
ADIT	AutoDep Input Tool	GEO	Gene Expression Omnibus
AE	annotated exon	GFP	green fluorescent protein
AN	actual negative	GGTC	German Gene Trap Consortium
AP	actual positive	GNOME	GNU network object model environment
AQL	ACeDB query language		
BDGP	Berkeley Drosophila Genome Project	GOLD	Genomes Online Database
BIND	Biomolecular Interaction Network Database	GOR	Garnier–Osguthorpe–Robson
		GRAIL	Gene Recognition and Assembly Internet Link
BIOS	Basic Input–Output System		
BRET	bioluminescent resonance energy transfer	GSS	genome survey sequence
		GST	glutathione S-transferase
CASP	Critical Assessment of Structure Prediction	GUI	graphical user interface
		HIV	human immunodeficiency virus
CD	circular dichroism	HMM	hidden Markov model
CD	candidate drug	HSP	high-scoring segment pair
CDE	common desktop environment	HTG	high-throughput genomic sequence
cDNA	copy DNA		
CDS	coding sequence	HTML	hypertext markup language
CGI	common gateway interface	HTS	high-throughput screening
CIP	Cahn–Inglod–Prelog	http	hypertext transfer protocol
CORBA	Common Object Request Brokering Architecture	IP	Internet Protocol
		ISP	Internet Service Provider
DBMS	database management system	KDE	K desktop environment
DDBJ	DNA Databank of Japan	KEGG	Kyoto Encyclopedia of Genes and Genomes
DEC	Digital Equipment Corporation		
DIP	Database of Interacting Proteins	LCA	last common ancestor
		LOG	Laplacian of Gaussian
DNS	Domain Name Server	m/e or m/z	mass/charge ratio
EBI	European Bioinformatics Institute	MAD	multiwavelength anomalous diffraction
EMBL	European Molecular Biology Laboratory	MAGE	microarray and gene expression
		MAGE-ML	microarray gene expression markup language
ENU	ethylnitrosourea		
EP	Expression Profiler	MAGE-OM	microarray gene expression object model
ES cells	Embryonic stem cells		
ESI	electrospray ionization	MALDI	matrix-assisted laser desorption/ionization
EST	expressed sequence tag		
ExPASy	Export Protein Analysis System (Switzerland)	ME	missing exon
		MGED	Microarray Gene Expression Database
FE	false exon		
FN	false negative	MIAME	minimum information about a microarray experiment
FP	false positive		

MIME	Multiple Internet Mail Extensions	RMSD	root mean square deviation
MIR	multiple isomorphous replacement	rRNA	ribosomal RNA
		RT	reverse transcription
MMDB	Molecular Modeling Database	SAGE	serial analysis of gene expression
mRNA	messenger RNA	SDS	sodium dodecyl sulfate
MS	mass spectrometry	SELDI	surface-enhance laser desorption/ionization
MSD	Macromolecular Structure Database	SH2, SH3	Src – homology domain
MS-DOS	Microsoft Disk Operating System	SMART	Simple Modular Architecture Research Tool
MSF	multiple sequence format	SMILES	Simplified Molecular Input Line Entry Specification
NBRF	National Biomedical Research Foundation	SNP	single nucleotide polymorphism
NCBI	National Center for Biotechnology Information	SOM	self-organizing map
		SPR	surface plasmon resonance
NDB	Nucleic Acid Data Bank	SQL	symbolic query language
NJ	neighbor joining	SRS	sequence retrieval system
NMR	nuclear magnetic resonance	SSE	secondary structure element
NNSSP	Nearest Neighbour Secondary Structure Prediction	STS	sequence tagged site
		T_C	Tanimoto coefficient
NOE	nuclear Overhauser effect	TCP	Transmission Control Protocol
OMIM	OnLine Mendelian Inheritance in Man	TE	true exon
		TN	true negative
ORF	open reading frame	TP	true positive
PAGE	polyacrylamide gel electrophoresis	TrEMBL	Translated EMBL
		tRNA	transfer RNA
PAM	accepted point mutations	UML	Unified Modeling Language
PAUP	phylogenetic analysis using parsimony	UPGMA	unweighted pair group method using arithmetic mean
PCNA	proliferating cell nuclear antigen	UPGMC	unweighted pair group method using centroid value
PCR	polymerase chain reaction		
PDB	Protein Data Bank	URL	uniform resource locator
PE	predicted exon	WE	wrong exon
PERL	Practical Extraction and Reporting Language	WPGMA	weighted pair group method using arithmetic mean
PH	pleckstrin homology	WPGMC	weighted pair group method using centroid value
PHYLIP	phylogenetic inference package		
pI	isoelectric point	WST	watershed transformation
PIR	Protein Information Resource	WWW	World Wide Web
PN	predicted negative	XML	eXtensible Markup Language
PP	predicted positive	Y2H	yeast two-hybrid
RCSB	Research Collaboratory for Structural Bioinformatics		

PREFACE

Computational analysis of biological sequences has been practised for a long time, but its importance to the practising biochemist or molecular biologist has only emerged in the last ten years or so as high-throughput sequencing techniques have brought a flood of valuable sequence data of relevance to many research projects. Thus the discipline of bioinformatics has emerged and grown enormously in importance. Center stage in these developments has been the sequencing of whole genomes. The human genome was finished more than a year ago and convenient access to the annotated data for the non-bioinformatician is just becoming available.

Following DNA sequencing, other experimental techniques have been developed to the point where they can be described as high-throughput. Gene- and protein-expression patterns can be studied with microarrays on the scale of whole genomes, protein interactions can be studied on the same scale with the yeast two hybrid system, and there is even talk of high-throughput structure determination. Where there is large-scale data generation there is need for computational data handling and analysis, and so the subject of bioinformatics grows.

It is now crucial that the practising experimental biologist has a knowledge of bioinformatics, and so the subject is beginning to appear and grow in undergraduate curricula. This book is essentially aimed at the experimental biologist, perhaps as an undergraduate or maybe further along a career path, who needs a working knowledge of the subject. It would also be usful to computer scientists and others looking to move into a biological domain. We begin by describing data generation and databases, and then move to the 'classical' bioinformatics question, "What can I do with my sequence?". Following this we look at the newer bioinformatics problems associated with structure, expression, proteomics, interactions and pathways.

Throughout the book we aim to describe what is possible, and the strengths, limitations and potential pitfalls of methods and analyses. We will tell you how to do things, but this is not a software manual for commonly used packages. They have their own manuals that are (mostly) much better than anything we could provide. Many of the methods we describe rely on quite complex mathematical, statistical or computational techniques. Often we choose not to describe these at all, but where we do we have aimed for a simple conceptual understanding. It is often said that it is not necessary to understand in detail how the internal combustion engine works in order to drive a car. The same applies to understanding of underlying methods in bioinformatics, but where a little understanding is possible and helpful, we have tried to provide it. We hope that it is useful.

Acknowledgments

The authors would like to thank Jonathan Ray, Sarah Carlson and David Hames for their help, understanding and support during the preparation of the book, and Chris Hodgson for advice and discussion on Section O.

David Westhead dedicates this book to his parents Robert and Mavis, his wife Andrea, and children Elizabeth and Francis.

Howard Parish dedicates this book to his grand-daughter Scarlett.

Richard Twyman dedicates this book to his parents, Peter and Irene, his children, Emily and Lucy, and to Hannah, Joshua and Dylan.

A1 THE SCOPE OF BIOINFORMATICS

Key Notes

What is bioinformatics?

Bioinformatics is the combination of biology and information technology. It is the branch of science that deals with the computer-based analysis of large biological data sets. Bioinformatics incorporates the development of databases to store and search data, and of statistical tools and algorithms to analyze and determine relationships between biological data sets, such as macromolecular sequences, structures, expression profiles and biochemical pathways.

The role of computers in bioinformatics

Computers are required in bioinformatics for their processing speed (allowing repetitive tasks to be carried out quickly and systematically) and for their problem-solving power. However, many problems addressed by bioinformatics still require expert human input, and the integrity and quality of the source data are also critical.

Scope of this book

This book is designed to provide the newcomer to bioinformatics with enough information to understand the principles of bioinformatic applications and to gain some practice in their use. The text covers basic introductory subjects such as the role of the Internet as well as the key areas of bioinformatics: the use of databases, sequence and structural analysis tools, and tools for annotation, expression analysis and the analysis of biochemical and molecular pathways. Section O is an appendix of peripheral information on computer operating systems and software.

Instant Notes Bioinformatics WWW site

Throughout this book there are references to various WWW sites that house information resources, databases and bioinformatic tools. Since the WWW is constantly evolving, the addresses of these sites are likely to change. For convenience, links to these sites are listed on an accompanying WWW site (http://www.bios.co.uk/inbioinformatics), which will be regularly updated by authors. If any of the URLs in this book do not work, the accompanying WWW site is probably the best way to reach the required site.

Related topics

Bioinformatics and the Internet (A2)

Useful bioinformatics sites on the WWW (A3)

What is bioinformatics?

Bioinformatics is the marriage of biology and information technology. The discipline encompasses any computational tools and methods used to manage, analyze and manipulate large sets of biological data. Essentially, bioinformatics has three components:

- The creation of **databases** allowing the storage and management of large biological data sets.
- The development of **algorithms** and **statistics** to determine relationships among members of large data sets.
- The use of these tools for the analysis and interpretation of various types of biological data, including DNA, RNA and protein sequences, protein structures, gene expression profiles and biochemical pathways.

The term bioinformatics first came into use in the 1990s and was originally synonymous with the management and analysis of DNA, RNA and protein sequence data. Computational tools for sequence analysis had been available since the 1960s but this was a minority interest until advances in sequencing technology (Topic B1) led to a rapid expansion in the number of stored sequences in databases such as GenBank (Topic C2). Now the term has expanded to incorporate many other types of biological data, for example protein structures, gene expression profiles and protein interactions. Each of these areas requires its own set of databases, algorithms and statistical methods, some of which are discussed in this book.

The role of computers in bioinformatics

Bioinformatics is largely although not exclusively a computer-based discipline. Computers are important in bioinformatics for two reasons. First, many bioinformatics problems require the same task to be repeated millions of times. For example, comparing a new sequence to every other sequence stored in a database (Topic E3) or comparing a group of sequences systematically to determine evolutionary relationships (Topic G2). In such cases, the ability of computers to process information and test alternative solutions rapidly is indispensable. Second, computers are required for their problem-solving power. Typical problems that might be addressed using bioinformatics could include solving the folding pathway of a protein given its amino acid sequence, or deducing a biochemical pathway given a collection of RNA expression profiles. Computers can help with such problems, but it is important to note that expert input and robust original data are also required.

Scope of this book

This book is based on the authors' experience in teaching bioinformatics at undergraduate and postgraduate level. A common starting point for those new to bioinformatics is 'What can I do with this sequence?' This book is designed to give the reader an informed background to understanding methods used in bioinformatics and sufficient examples and technical details to enable him or her to answer real problems. We describe the role of the Internet in bioinformatics (Section A), how data used in bioinformatics is generated (Section B), the importance of databases (Section C) and how these are accessed and searched (Section D). We discuss sequence analysis (Sections E, F and G), sequence annotation (Section H), structural analysis and prediction (Section I), gene and protein expression analysis (Sections J and K) the bioinformatics of protein interactions (Sections L and M) and some applications of bioinformatics in the pharmaceutical industry (Section N). Section O comprises a series of appendices providing background information on file formats, computer operating systems and software.

There are topics in computational biology that we have intentionally omitted. Software designed specifically for structure refinement, automated instrumentation (including robotics) and other types of data collection are omitted. We include methods for molecular graphics but, otherwise, we omit graphical and other aids to document presentation.

Instant Notes Bioinformatics WWW site   This book makes reference to many databases and computer software tools that are available on the World Wide Web as well as various informative web sites. Although the addresses for many of these resources are listed in the book, the Internet is constantly evolving and such addresses are subject to change on a regular basis. For convenience, links to all the sites discussed in the text can be found on an accompanying WWW site (http://www.bios.co.uk/inbioinformatics). The WWW site also contains further information, updates and links that are not found in this book.

A2 BIOINFORMATICS AND THE INTERNET

Key Notes

The Internet	The Internet is an international computer network that uses a particular protocol, known as TCP/IP, to package and route data. Most academic, government and commercial institutions have access to the Internet, and individuals may also gain access by subscribing to an ISP. The Internet provides a versatile system for exchanging biological data.
The World Wide Web (WWW)	The WWW is an Internet-based system for information exchange using a protocol called http. Programs called browsers can access hypertext documents on the Internet by searching for the relevant address, known as a URL. Hypertext documents contain text and other multimedia objects (images, audio files, etc.) but their most important property is the presence of hyperlinks that allow the browser direct access to other hypertext documents. These may be hosted by any computer on the Internet, and in this way, the user can rapidly jump from computer to computer around the world viewing and downloading information.
Browsing, working and downloading	The WWW is a rich source of biological data and bioinformatics resources. Many bioinformatic tools and databases can be accessed and used over the Internet. However, due to constraints in the speed of data transfer or access to the Internet, it may also be appropriate to download databases and bioinformatic software and use them on a local computer.
Related topics	Useful bioinformatics sites on the WWW (A3) Installing bioinformatic software locally (O4)

The Internet

Biological information is stored on many different computers around the world. The easiest way to access this information is for the computers to be joined together in a network. A **computer network** is a group of computers that can communicate, for example over a telephone system, therefore allowing data to be exchanged between remote users. A typical computer network is shown in *Fig. 1*. For transfer, data are first broken into small **packets** (units of information), which are sent independently and reassembled when they arrive at their destination. If information is sent from computer A to computer C, it can travel

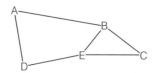

Fig. 1. A simple computer network.

via two different routes. In one case computer B acts as a **router**, and in the other case computers D and E both act as routers. The availability of different routes through the network means that communications can be maintained between computers A and C even if part of the network is unavailable, for example if computer B ceases to function.

The **Internet** is an international network of computers derived from an earlier system, **ARPAnet**, developed by the US military. The Internet as we know it began in 1969, when four American universities were connected together for the first time allowing the rapid exchange of scientific data. The number of computers linked to the Internet has grown exponentially over the last 30 years and it is now estimated that over 20 million computers have access, many of them personal computers in people's homes. Information transfer over the Internet is governed by a set of **protocols** (procedures for handling data packages) called **TCP/IP**. TCP is the **Transmission Control Protocol**, which determines how data is broken into packages and reassembled. IP is the **Internet Protocol**, which determines how the packets of information are addressed and routed over the network. To access the Internet, a computer must have the correct hardware (generally a network card and/or a modem), the appropriate software and permission for network access. Many institutions have automatic access to the Internet, but private users must subscribe to an **Internet Service Provider (ISP)**.

The World Wide Web (WWW)

The **World Wide Web (WWW)** is a way of exchanging information over the Internet using a program called a **browser**. A number of browsers are available for working on the WWW, the most widely used of which are **Internet Explorer** and **Netscape Navigator**. Most computers are sold nowadays 'Internet ready' with the appropriate hardware and one or both of these browser programs installed as standard. The WWW was developed in 1992 and allows the display of information pages containing **multimedia objects** (e.g. text, images, audio and video) in a special format called **hypertext**. In a hypertext document, text is displayed normally and can be read and manipulated like any other text document, but some words and objects are highlighted in a different color and these are known as **hypertext links** (or simply **hyperlinks**). Clicking on a hyperlink directs the browser to access another hypertext document, which might be on the same computer or might be on any other computer linked to the Internet. The new document may have its own hyperlinks and thus the process can be repeated allowing the user to move rapidly from computer to computer around the world downloading information as he or she goes (this is commonly known as **surfing the web** or **surfing the net**).

The WWW works on the basis that each hypertext document has a unique address known as a **uniform resource locator (URL)**. URLs take the format http://restofaddress, where 'http://' identifies the protocol for communication over the WWW (**hypertext transfer protocol**) and 'restofaddress' provides a location for the hypertext document on the Internet. Every computer on the Internet has an **IP address**, which is in the form of four integers conventionally separated by dots. Associated with this is a text version of the address, for example http://www.bios.co.uk, which is the publisher's address. The equivalent IP address for the publisher is 195.172.6.15. If a local user tries to contact http://www.bios.co.uk, how does the browser find the correct site? The local computer first contacts Internet computers called **Domain Name Servers (DNSs)** that try to understand parts of the address starting with the most significant (right hand) part. For example, most text addresses have a country abbreviation, in this case 'uk' for United Kingdom, but American addresses do not since the Internet was an American invention. If the computer one is trying to access is providing a service on the WWW, it is known as a **web server**. This means there are numerous files available for browsing, and each can be identified by a unique URL. Such files are specified by extra characters separated from the main Internet address by a solidus (/). For example, the URL http://www.bios.co.uk/bioinformatics refers to a subdirectory on the publisher's web server that corresponds to the web site accompanying this book. Once the DNS has found the Internet name for the server, it is for the server itself to work out what do about any extensions to the URL such as '/bioinformatics'.

Browsing, working and downloading

Browsing the Internet is simply a case of clicking on the desired hyperlinks and allowing the associated pages to download. Some pages download faster than others, which may be due to content (pages with many images and other large files will take longer to download than pages that contain text alone) or due to the speed of connection (there are bottlenecks in many parts of the Internet, and it is advisable to find a local web server to minimize the number of routers the information has to pass through). It is also notable that the Internet will be busier at certain times of the day, and during the weekends when recreational use increases. Many bioinformatics sites are hosted by several web servers in different locations around the world to reduce such bottlenecks. Different web servers providing the same service are called **mirrors**.

To access a particular web site, it may first be necessary to type in the URL in the **address bar** of the browser. Once a page has been accessed, however, it should not be necessary to type in the URL again. Browser programs maintain a list of URLs that have been visited (the **History file**) and any URL can be added to a list of **Favorites** (in Internet Explorer) or **Bookmarks** (in Netscape Navigator) to allow easy access in the future. Where does one start on the Internet? A number of public **search engines** are available allowing the user to search for sites of interest using particular keywords, but it may be better to start with some dedicated bioinformatics sites. For the absolute beginner, a selection of useful bioinformatics web sites is listed in Topic A3.

Having got the feel of bioinformatics on the WWW, what are the merits and demerits of installing software locally (Topic O4), rather than using a WWW site? Although locally installed software will usually run faster than the same application used over the Internet, some software is difficult to install and might need expert help. There are advantages in having local copies of simple sequence alignment and other software if you are working 'at home', that is, limited by rates of data transmission on telephone lines. However, the use of locally installed databases can be disadvantageous because updates will be published less frequently than the WWW-based versions. Many academic institutions have an **Intranet**, that is, a local network that can be accessed only from computers within the institution. Such local networks may provide a number of bioinformatics tools and applications, which will usually run just as fast as locally installed software.

A3 USEFUL BIOINFORMATICS SITES ON THE WWW

Key Notes

Useful bioinformatics sites	For the absolute beginner, the Internet can be daunting and intimidating. Nine good starting points (gateways) for bioinformatics on the Internet are listed. Each of the sites is well maintained, simple to use and provides a wealth of resources such as links, databases and bioinformatics software.
Searching the Internet	Once the user has gained some experience using the Internet, information can be found relatively easily. If the bioinformatic gateways do not provide the information required, general-purpose search engines can be used to locate pages containing specific key words or phrases. Otherwise, the home pages of academic institutions and biotechnology companies can be useful starting points.
Pitfalls and hints	When using a search engine it is important to refine the search and to avoid ambiguous words and phrases, as these can pull out numerous irrelevant pages. Literature databases can help by providing useful keywords and phrases to use as search terms.
Related topics	Bioinformatics and the Internet (A2) Annotated sequence databases (C2)	Genome and organism-specific (C3) Miscellaneous databases (C4)

Useful bioinformatics sites

For absolute beginners, we have listed nine good starting points for bioinformatics on the WWW (*Table 1*). Each of these **gateway sites** is comprehensive, has many useful links and is well maintained and stable. Time spent browsing and using these sites will provide an accurate feeling for the bioinformatics resources available on the Internet.

Searching the Internet

Although the nine web sites listed in *Table 1* provide some of the best starting points for bioinformatics on the WWW, there is a great deal of specialist biological data that cannot be accessed directly from these sites. Finding relevant data on the Internet is made simpler by the availability of general-purpose **search engines,** such as Google, Yahoo, Lycos, AltaVista and Hotbot. These tools search the entire Internet for pages that contain particular keywords or phrases, and they can also be used to search for files of a particular type, such as image files or video files. For example, one might search the Internet using the phrase 'alcohol dehydrogenase' to find pages containing information about that enzyme. Alternatively, one might look for image files of a particular insect or flower, or video files of frog development. Relevant sites are displayed as a list

Table 1. Nine good starting points for bioinformatics on the WWW

URL	Note
General bioinformatics 'gateways'	
http://www.ncbi.nlm.nih.gov/	National Center for Biotechnology Information homepage. A resource for public databases, bioinformatics tools and applications. Links to many useful sites and resources for bioinformatics software.
http://www.ebi.ac.uk/	The EMBL European Bioinformatics Institute outstation. A resource for biological databases and software, much of which has excellent tutorial support.
http://www.expasy.ch/	The ExPASy (Expert Protein Analysis System) Molecular Biology Server. Maintained by the Swiss Institute of Bioinformatics (SIB). Provides links, databases and software resources for the analysis of protein sequences, structures and expression.
http://www.embl-heidelberg.de/	European Molecular Biology Laboratory homepage
http://www.gmd.de/Welcome.en.html	German National Centre for Information Technology homepage
http://links.bmn.com/	The BioMedNet gateway to thousands of biological websites, includes a search facility and provides descriptions of each of the web sites listed.
Genome projects	
http://wit.integratedgenomics.com/GOLD/	Genomes On Line Database, with links to genomic databases and progress reports on genome projects.
http://www.genome.ad.jp/kegg/	Kyoto Encyclopaedia of Genes and Genomes. A very comprehensive Japanese site including metabolic maps.
Computing notes	
http://foldoc.doc.ic.ac.uk/foldoc/index.html	FOLD (Free Online Dictionary of Computing). A good place to look up meanings of computer jargon.

of **hits**, with hyperlinks allowing direct access to the page of interest. The problem with general-purpose search engines is that they have not been developed specifically with molecular biology in mind, and the information they provide can be irrelevant or misleading, especially if the search term used has other connotations.

As an alternative to search engines, the home pages of academic institutions or biotechnology companies can also be a good place to start. Many universities, for example, maintain comprehensive web sites with pages for staff to describe research projects and display data, and such sites often contain hyperlinks to sites of related interest.

Pitfalls and hints

On a general-purpose search engine it is probably better to start with a set of key words that is very restrictive and then remove some of the words if no hits are generated. If a search term is too broad (e.g. 'biochemistry') it will produce a ridiculous number of hits and it will be impossible to check all the listed pages. Search terms with known alternative uses are also best avoided. For example, searching the Internet with the word 'steroid' will likely hit more pages on body-building than molecular biology! A positive suggestion is to use a literature database on the WWW such as PubMed (Topic C4) to look for useful and appropriate keywords and phrases to use as search terms.

B1 SEQUENCING DNA, RNA AND PROTEINS

Key Notes

Principles of DNA sequencing

DNA sequencing is performed using an automated version of the chain termination reaction, in which limiting amounts of dideoxyribonucleotides generate nested sets of DNA fragments with specific terminal bases. Four reactions are set up, one for each of the four bases in DNA, each incorporating a different fluorescent label. The DNA fragments are separated by PAGE and the sequence is read by a scanner as each fragment moves to the bottom of the gel.

Types of DNA sequence

DNA sequences come in three major forms. Genomic DNA comes directly from the genome and includes extragenic material as well as genes. In eukaryotes, genomic DNA contains introns. cDNA is reverse-transcribed from mRNA and corresponds only to the expressed parts of the genome. It does not contain introns. Finally, recombinant DNA comes from the laboratory and comprises artificial DNA molecules such as cloning vectors.

Genome sequencing strategies

Only short DNA molecules (~800 bp) can be sequenced in one read, so large DNA molecules, such as genomes, must first be broken into fragments. Genome sequencing can be approached in two ways. Shotgun sequencing involves the generation of random DNA fragments, which are sequenced in large numbers to provide genome-wide coverage. Conversely, clone contig sequencing involves the systematic production and sequencing of subclones.

Sequence quality control

High-quality sequence data is generated by performing multiple reads on both DNA strands. The preliminary trace data is then base called and assessed for quality using a program such as Phred. Vector sequences and repeated DNA elements are masked off and then the sequence is assembled into contigs using a program such as Phrap. Remaining inconsistencies must be addressed by human curators.

Single-pass sequencing

Sequence data of lower quality can be generated by single reads (single-pass sequencing). Although somewhat inaccurate, single-pass sequences such as ESTs and GSSs can be generated in large amounts very quickly and inexpensively.

RNA sequencing

Most RNA sequences are deduced from the corresponding DNA sequences but special methods are required for the identification of modified nucleotides. These include biochemical assays, NMR spectroscopy and MS.

Protein sequencing	Most protein sequencing is nowadays carried out by MS, a technique in which accurate molecular masses are calculated from the mass/charge ratio of ions in a vacuum. Soft ionization methods allow MS analysis of large macromolecules such as proteins. Sequences can be deduced by comparing the masses of tryptic peptide fragments to those predicted from virtual digests of proteins in databases. Also, de novo sequencing can be carried out by generating nested sets of peptide fragments in a collision cell and calculating the difference in mass between fragments differing in length by a single amino acid residue.
Related topics	Sequence similarity searches Principles of genome annotation (E1) (H1)

Principles of DNA sequencing

The order of nucleotides in DNA is determined by **chain termination sequencing** (also called **dideoxy sequencing** or the **Sanger method** after its inventor). The basic sequencing reaction, which is summarized in *Fig. 1*, consists of a **single-stranded DNA template**, a **primer** to initiate the nascent chain, four **deoxyribonucleoside triphosphates** (dATP, dCTP, dGTP and dTTP) and the enzyme **DNA polymerase**, which inserts the complementary nucleotides in the nascent DNA strand using the template as a guide. Four DNA polymerase reactions are set up, each containing a small amount of one of four **dideoxyribonucleoside triphosphates** (ddATP, ddCTP, ddGTP or ddTTP). These act as chain-terminating competitive inhibitors of the reaction, but are present in limiting amounts so there is only a small chance the growing chain will terminate at any given position. Therefore, each of the four reaction mixtures generates a **nested set** of DNA fragments, each terminating at a specific base.

Nowadays, most DNA sequencing reactions are **automated**. Each reaction mixture is labeled with a different fluorescent tag (on either the primer or on one of the nucleotide substrates), which allows the terminal base of each fragment to be identified by a scanner. All four reaction mixtures are then pooled and the DNA fragments separated by polyacrylamide gel electrophoresis (PAGE). Smaller DNA fragments travel through the gel faster than larger ones so the nested DNA fragments are separated according to size. The resolution of PAGE allows polynucleotides differing in length by only one residue to be separated. Near the bottom of the gel, the scanner 'reads' the fluorescent tag as each DNA fragment moves past, and this is converted into **trace data**, displayed as a graph comprising colored peaks corresponding to each base (*Fig. 2*).

Types of DNA sequence

DNA sequences are stored in databases such as GenBank (Topic C2) and are generally of three types: genomic DNA, cDNA and recombinant DNA. **Genomic DNA** is taken direct from the genome and contains genes in their natural state, which in eukaryotes includes introns, regulatory elements and large amounts of surrounding intergenic DNA. In contrast, **copy DNA (cDNA)** is prepared by reverse transcribing messenger RNA (mRNA). It is often useful to focus on cDNA because this provides direct access to genes, which may be difficult to find in genomic DNA. In the human genome for example, genes represent only 3% of the sequence. **Recombinant DNA** includes the sequences of vectors such as plasmids, modified viruses and other genetic elements that are used in the laboratory. It is essential for these sequences to be stored in

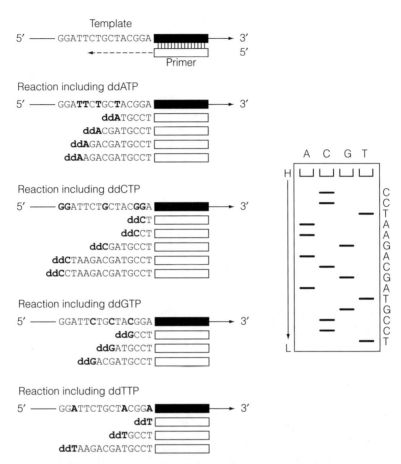

Fig. 1. Principle of DNA sequencing. (a) Four sequencing reactions are set up, each containing a limiting amount of one of the four dideoxynucleotides. Each reaction generates a nested set of fragments terminating with a specific base as shown. (b) A polyacrylamide gel is shown with each reaction running in a separate lane for clarity. In a typical automated reaction, all reactions would be pooled prior to electrophoresis and the terminal nucleotide determined by scanning for a specific fluorescent tag. From Twyman, R.M., Advanced Molecular Biology, © BIOS Scientific Publishers Limited, 1998.

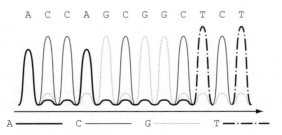

Fig. 2. A sample of a high-quality sequence trace, where all peaks are easily called. Peaks are typically printed in different colors (shown here as different line styles) to aid visual interpretation. Software such as Phred is used to read the peaks and assign quality values (http://www.phrap.org).

databases because linked vector sequences must be removed from new genomic and cDNA sequences prior to alignment to avoid spurious matches.

Genome sequencing strategies

Using current technology, only about 800 nucleotides of DNA sequence can be read in a single reaction. If a larger DNA molecule, such as an entire genome, is to be sequenced, it must first be broken up into smaller pieces, each of which has to be sequenced individually. There are two strategies for assembling genome sequences from short reads. In the **shotgun sequencing** strategy, the large DNA molecule is randomly broken up and many sequencing reactions are performed to cover the entire molecule. The entire sequence is reassembled by using a computer to search for overlaps. This method generates a large amount of sequence data very quickly, but difficulty may be experienced in closing the final gaps, a process termed **finishing**. In the alternative **clone contig** approach, DNA fragments are subcloned in a rational manner and systematically sequenced until the entire sequence is completed. Using this method, sequence data accumulates more slowly at the beginning of a project, but there are fewer sequence gaps at the end so finishing is easier.

Sequence quality control

Genome sequencing is subject to strict **quality control** measures. For each clone, the sequencing reaction is performed several times on each strand. This primary trace data is then subject to **pre-assembly processing**, which includes base calling and quality assessment, the removal of vector sequences and the masking of repeats. Finally, the sequence reads are assembled into **contigs** (contiguous sequence reads). **Base calling** means deciding to which base a particular trace peak corresponds. Ideally, each peak is clear and unambiguous (*Fig. 2*), but the quality of trace data can vary considerably. Several computer programs are available that automatically make base calls from sequence traces, and assess the quality (i.e. confidence) of each call. One of the most widely used utilities is **Phred**, which uses dynamic programming (Topic E1) to match observed and predicted peak locations and evaluate the quality of the trace data for each base at each position. The next processing step is to remove vector sequences. Sequence reads usually start with part of the sequencing vector and it is important to remove this sequence because it can generate false overlaps. Programs such as **vector_clip** and **CrossMatch** have been developed which align new sequence reads to a database of known vector sequences, and then mask off the vector sequence from further analysis. An essential step in genome sequence assembly is the identification of repetitive DNA elements. Genomic DNA is rich in repetitive elements, which can be arranged in tandem or dispersed throughout the genome. Such repeats are extremely troublesome when it comes to reassemble individual reads into contigs because they generate false overlaps. The program **RepeatMasker** is widely used to identify and tag such elements. Finally, the assembly process itself, which is carried out using programs such as **Phrap**. This is a computationally difficult task owing to the large number of possible overlaps. For example, if there are 100 reads, there are nearly 40 000 different arrangements. The Phrap assembly method is iterative and based on calculating the log likelihood ratio for each possible match taking into account both the sequence and the trace quality data. Despite the high level of automation, human curation to resolve disputed base calls and other discrepancies is required. Editing and contig assembly can be carried out using programs such as **Staden Gap4**, which allow trace data to be evaluated and the results of multiple sequence reads to be displayed.

Single-pass sequencing

As discussed above, multiple sequencing reactions on both strands of a cloned DNA molecule are usually required to insure accuracy. However, where large amounts of sequence data are generated (e.g. in genome projects) **single-pass sequencing** is quicker, cheaper and provides more data at the expense of some accuracy. GenBank has subsidiary databases for such sequences. For example, **dbEST** is a database for **expressed sequence tags (ESTs)**, which are generated by rapid, single-pass sequencing of random clones from cDNA libraries. Although short (2–300 bp) and inaccurate, very large numbers of sequences have been collected rapidly and inexpensively, and can be used to identify genes in genomic DNA and to prepare large clone sets for DNA microarrays (Topic B3). The genomic equivalent of dbEST is **dbGSS**, a database for **genome survey sequences (GSSs)**, which are random, single-pass sequences of genomic DNA. Large amounts of preliminary genomic sequence data (**unfinished sequence**) are stored in the **HTG (high-throughput genomic)** division of GenBank.

Note that the vast majority of ESTs are eukaryotic sequences. This is because cDNA synthesis is generally carried out using a primer that anneals to the **polyadenylate tail** of the eukaryotic mRNA. Most bacterial mRNAs lack a polyadenylate tail and cannot be copied using the same strategy. Furthermore, many bacterial transcripts are never synthesized as full-length molecules due to their operon organization and the coupling of transcription and translation. However, bacteria also tend to have compact genomes and genes that lack introns, therefore genomic DNA fragments are already very similar to cDNAs.

RNA sequencing

RNA sequencing is less straighttorward than DNA sequencing since there are large numbers of **minor nucleotides** (chemically modified nucleotides), for example in transfer RNA (tRNA) and ribosomal RNA (rRNA). An RNA sequence lacking data on minor nucleotides can be deduced from either the DNA sequence of the corresponding gene or from a cDNA sequence (the sequence of the reverse transcript). In many cases, cDNA sequences are more informative than gene sequences because the RNA molecule may be extensively processed during synthesis. For example, **introns** may be spliced out of a primary transcript to generate a mature mRNA, and extra nucleotides may be inserted by **RNA editing**. Direct RNA sequencing involves the chemical characterization of modified nucleotides. One type of modification, 2′-O-methylation, is easy to pick up as the phosphodiester bond on the 3′-side of such a nucleotide is not alkali-labile. Other types of modification can be identified by mass spectrometry (see below).

Protein sequencing

Protein sequences can be deduced from DNA sequences, but as with RNA sequencing this does not provide information on modified residues or other types of post-translational protein modification (such as cleavage or the formation of disulfide bonds). In the past, direct protein sequencing was carried out using a process called **Edman degradation**, in which the terminal residue of a protein was labeled, removed and then identified using a series of chemical tests. Current methods for protein sequencing rely on **mass spectrometry (MS)**, a technique in which the **mass/charge ratio** (m/e or m/z) of ions in a vacuum is accurately determined, allowing molecular masses to be calculated at a resolution in the order of 10^{-5}. Only extremely small quantities of material are required for analysis and the resolution of the technique is such that, for small molecules, the accurate m/e value directly reveals the chemical formula. Hence, the technique can be used to characterize minor nucleotides in nucleic acids (see above).

MS of macromolecules such as proteins relies on so-called **soft ionization** methods, which allow then ionization of large molecules without degradation. Two methods are commonly used: **ESI (electrospray ionization)** and **MALDI (matrix-assisted laser desorption/ionization),** and these are discussed in more detail in Section K. The simplest application of MS is to catalog the m/e values for the proteolytic (usually tryptic) digests of a protein. These m/e values are then compared with either a database of the predicted tryptic digests of proteins or, given a suitable database of protein sequences, with 'virtual digests' performed in real time on a computer. However, this requires that the protein in question already be characterized. For the complete sequencing of an unknown peptide by MS, the protein can be randomly fragmented in a **collision cell** within the mass spectrometer, generating a nested set of fragments. Given a knowledge of the molecular masses of all amino acids, calculating the m/e ratios of all these fragments (and more specifically, the differences between them) allows the sequence to be deduced. In reality, the problem is complicated by the fact there are two series of fragments generated at the same time: these are known as the **B series** (*N*-terminal fragments) and the **Y series** (*C*-terminal fragments). The principle is summarized in *Fig. 3*. The problem can be addressed by labeling the protein with, for example, an *N*-terminal tag, which would add recognized mass to all fragments in the B series. Note that the amino acids leucine and isoleucine cannot be distinguished by MS because they have the same relative molecular mass.

Y series

```
                        <-------------- Y2 --------------->
                                     <-------Y1 ------->
        H₂NCH(R¹H)C(=O)---NHCH(R²H)C(=O)--- NHCH(R³H)C(=O)OH

            <---- B1 ---->
            <------------ B2 ----------->
```

B series

Fig. 3. Illustration of the origin of B and Y series ions in the fragmentation of a hypothetical tripeptide. Peptide bonds are shown as lines. B1 and Y1 are the two smallest fragments (single amino acids) and B2 and Y2 are the dipeptides. In general, a peptide of N residues contains (N–1) peptide bonds and, assuming the collision cell only breaks one peptide bond in each molecule, the Y and B series will each contain (N–1) peptides.

B2 DETERMINATION OF PROTEIN STRUCTURE

Key Notes

X-ray crystallography
X-ray crystallography involves the determination of protein structure by studying the diffraction pattern of X-rays through a precisely orientated protein crystal. The way in which X-rays are scattered depends on the electron density and spatial orientation of the atoms in the crystal. A mathematical method called the Fourier transform is used to reconstruct electron density maps from the diffraction data allowing structural models to be built.

NMR spectroscopy
NMR is a property of certain atoms that can switch between magnetic states in an applied magnetic field by absorbing electromagnetic radiation. The nature of the absorbance spectrum is influenced by the type of atom and its chemical context, so that NMR spectroscopy can discriminate between different chemical groups. NMR spectra are also modified by the proximity of atoms in space. Analysis of NMR spectra therefore allows the three-dimensional configuration of atoms to be reconstructed, resulting in a series of structural models. The technique is suitable only for the analysis of small, soluble proteins.

Other methods
For larger proteins that do not readily form crystals, alternative analytical methods are required to deduce structures. These include X-ray fiber diffraction, electron microscopy and CD spectroscopy.

Related topics
Conceptual models of protein structure (I1)

Obtaining, viewing and analyzing structural data (I4)
Structural alignment (I5)

X-ray crystallography
The determination of protein structure by **X-ray crystallography** involves the reconstruction of atomic positions based on the diffraction pattern of **X-rays** through a precisely orientated **protein crystal**. Scattered X-rays cause positive and negative interference, generating an ordered pattern of signals called **reflections**. The amplitude and phase of the scattering depends on the number of electrons in each atom, allowing **electron density maps** to be reconstructed. This requires a mathematical technique called the **Fourier transform**, and the more data used the greater the resolution of the resulting structural model.

Structural determination depends on three variables: the **amplitude** and **phase** of the scattering and the **wavelength** of the incident X-rays. The wavelength of the X-rays is known and the amplitude of the scattering can be calculated from the intensities of the diffraction pattern. The phase of scattering, however, cannot be calculated directly. Phase can be determined by a number of indirect strategies including **multiple isomorphous replacement (MIR)** (the use of heavy atom derivatives of the protein to cause anomalous scattering) and multiwavelength anomalous diffraction (MAD), which uses synchrotron sources to provide X-rays with a range of incident wavelengths. X-ray crystallography used to be a laborious technique, but recent developments in protein expression, robotic workstations that can process hundreds of thousands of crystallization experiments in parallel, and high-throughput MAD analysis allow the relevant data to be collected in only a few weeks.

NMR spectroscopy

NMR (nuclear magnetic resonance) spectroscopy is used to determine the structure of small proteins (less than 25 kDa) in solution, and is often applied to proteins that do not readily form crystals. The basis of the technique is that some atoms, including natural isotopes of nitrogen, phosphorus and hydrogen, behave as tiny magnets and can switch between magnetic **spin states** in an applied magnetic field. This is achieved by the absorbance of low wavelength electromagnetic radiation (radio waves), generating **NMR spectra**. The resonance frequency for different atoms is unique, and is also influenced by the surrounding electron density. Thus, the magnetic resonance changes not only for different atoms but also in the context of different chemical groups (this is termed a **chemical shift**), allowing the discrimination between, for example, aryl and aromatic groups. The decay of nuclear magnetic resonance depends on the structure and spatial configuration of the molecule. The **nuclear Overhauser effect (NOE)** results from the transfer of magnetic energy through space if interacting nuclei are less than 0.5 nm apart and results in the appearance of symmetrical peaks superimposed over the normal one-dimensional NMR spectrum. The analysis of NMR spectra thus allows a set of **distance constraints** to be built up, which can be used to determine the three-dimensional spatial arrangement of atoms.

Other methods

X-ray crystallography and NMR spectroscopy account for the vast majority of three-dimensional structural data available for proteins. However, some proteins are both difficult to crystallize and too large for NMR. In such cases, other spectroscopic and related methods are required. For large integral membrane proteins, which are generally difficult to crystallize, it is often necessary to study the membrane as a whole. A modification of NMR – **magic angle spinning NMR** – allows sharp NMR signals to be obtained from preparations in films. The regular spacing of proteins in membranes also permits alternative techniques to be used, such as **electron microscopy**. The electron micrograph is used as a diffraction grating. In this way, structural information can be derived but, to date, with a lower resolution than other methods. Alternatively, the membrane with its proteins in place can be used as a two-dimensional crystal, and diffraction patterns can be obtained from X-rays with a low angle of incidence on the preparation. This method can also be used to determine the structure of fibrous proteins such as collagen and keratin, and is known as **fiber diffraction**. A useful complement to X-ray crystallography and NMR spectroscopy is **circular dichroism (CD) spectroscopy**, which exploits the optical activity of asymmetric molecules as shown by their differing absorption spectra in left and right circularly polarized light. CD spectroscopy between 160 and 240 nm can provide information about protein secondary structure, since proteins containing predominantly α-helices and proteins consisting mainly of β-sheets generate distinct spectra.

B3 GENE AND PROTEIN EXPRESSION DATA

Key Notes

Global expression analysis

Global expression analysis refers to any experiment in which the expression of all genes is monitored simultaneously. Such experiments generate large amounts of data, but unlike sequence and structural data, there is no universal system for the description of gene expression profiles. At the RNA level, expression data may be obtained as digital expression readouts following direct sequence sampling from libraries or databases, or using more sophisticated techniques such as SAGE. Most global RNA expression data, however, are obtained as signal intensities from microarray experiments. Global protein expression data are obtained predominantly as signal intensities on 2D protein gels.

DNA microarrays

DNA microarrays are the most widely used system for monitoring global gene expression at the RNA level. A microarray comprises a series of DNA elements (features) arranged in a grid pattern on a miniature support such as a glass chip. Hybridization with a complex probe allows the expression levels of many genes to be visualized simultaneously. If two probes are used, each labeled with a different fluorophore, differential gene expression between samples can be monitored directly on the same array. Two major technologies are used: spotted DNA microarrays and oligonucleotide GeneChips. In the latter case, the oligonucleotides are printed onto the chips using solid-state chemical synthesis during chip manufacture.

2D protein gels

Global protein expression analysis is achieved using high-resolution 2D gel electrophoresis. In this technique, proteins are separated in the first dimension by isoelectric focusing in an immobilized pH gradient, and in the second dimension according to molecular mass. After staining the gel, the resulting pattern of spots is a reproducible fingerprint of proteins in the sample. Comparison between samples can identify proteins that are differentially expressed, or induced in response to drugs, and so on. Excised spots are analyzed by mass spectrometry to characterize the proteins.

Related topics

Microarray data: analysis methods (J1)

Microarray data: tools and resources (J2)

Sequence sampling and SAGE (J3)

Analyzing data from 2D-PAGE gels (K1)

Analyzing protein mass spectrometry data (K2)

Global expression analysis

Gene expression can be studied at either the RNA or protein level. Traditionally, expression has been studied on a gene-by-gene basis using techniques such as Northern and Western blots. More recently, methods have been developed for **global expression analysis**, that is, the study of all genes simultaneously. Such studies generate very large amounts of data, which must be mined for relevant information using bioinformatics tools, but what forms does this data take?

For analysis at the RNA level, **direct sequence sampling** from RNA populations or cDNA libraries, or even from sequence databases, can be useful. A simple approach is to sequence say 5000 randomly picked clones from a cDNA library. Abundant mRNAs would appear at a higher frequency among the sampled sequences than rare ones, and statistical analysis of these data would allow relative expression levels to be determined. Although simple in concept, this type of experiment is expensive because of all the sequencing reactions that must be carried out. A more sophisticated technique is **serial analysis of gene expression (SAGE).** In this method, very short sequence tags (usually 8–15 nt) are generated from each cDNA and hundreds of these are joined together to form a concatemer prior to sequencing. Thus, in one sequencing reaction, information on the abundance of hundreds of mRNAs can be gathered. Each **SAGE tag** uniquely identifies a particular gene, and by counting the tags, the relative expression level of each gene can be determined. The advantage of all direct sampling methods is that expression levels are automatically converted into numbers, that is, the data are digital.

Although direct sampling is a powerful approach to expression analysis, it is labor intensive. The technology that has had the widest impact on global RNA expression profiling is hybridization to **DNA arrays (DNA chips)**. As discussed in more detail below, the principle advantage of DNA arrays is that the expression levels of all genes can be monitored in parallel in a single experiment. Expression data are obtained as signal intensities, and these are clustered to identify similarly expressed genes (Section J).

Array-based assay formats have also been developed for proteins but these 'protein chips' are not yet as versatile as the equivalent DNA chips. Currently, the most widely used technology for protein expression analysis is separation by **two-dimensional polyacrylamide gel electrophoresis (2D-PAGE)**, generating a unique pattern of dots (each dot representing a single protein). In 2D-PAGE experiments, protein expression data are obtained as signal intensities for each of the dots. However, unlike microarray experiments where each signal can be linked to its corresponding gene, each protein signal on a 2D gel has to be individually annotated by **mass spectrometry** (Section K).

DNA microarrays

Nucleic acids (DNA and RNA) can form double-stranded molecules by **hybridization**, that is, complementary base pairing. The specificity of nucleic acid hybridization is such that a particular DNA or RNA molecule can be labeled (e.g. with a radioactive or fluorescent tag) to generate a **probe**, and can be used to isolate a complementary molecule from a very complex mixture, such as whole genomic DNA or whole cellular RNA. This specificity also allows thousands of hybridization reactions to be carried out simultaneously in the same experiment.

A **DNA microarray** or **DNA chip** is a dense grid of DNA elements (often called **features** or **cells**) arranged on a miniature support, such as a nylon filter or glass slide. Each feature represents a different gene. The array is usually hybridized with a **complex RNA probe**, that is, a probe generated by labeling a complex mixture of RNA molecules derived from a particular cell type. The composition of such a probe reflects the levels of individual RNA molecules in its source. If nonsaturating hybridization is carried out, the intensity of the signal for each feature on the microarray represents the level of the corresponding RNA in the probe, thus allowing the relative expression levels of thousands of genes to be visualized simultaneously.

Two major technologies are used in microarray manufacture. The cheaper and currently more widely accessible method involves the robotic spotting of individual DNA clones onto a coated glass slide. Such **spotted DNA arrays** can have a density of up to 5000 features per square cm. The features comprise double-stranded DNA molecules (genomic clones or cDNAs) up to 400 bp in length and must be denatured prior to hybridization. Prefabricated arrays can be purchased from a number of companies and many laboratories have their own in-house production facilities. The alternative method is **on-chip photolithographic synthesis**, in which short oligonucleotides are synthesized *in situ* during chip manufacture. These arrays are known as **GeneChips** and are manufactured exclusively by the US company Affymetrix Inc. They have a density of up to 1 000 000 features per square cm, each feature comprising up to 10^9 single-stranded oligonucleotides 25 nt in length. Due to the reduced hybridization specificity of a short oligonucleotide, each gene on a GeneChip is represented by 20 features (20 nonoverlapping oligos), and 20 mismatching controls are included to normalize for nonspecific hybridization.

Both technologies can be used to look at differential gene expression between samples (e.g. healthy tissue vs. disease tissue). Fluorescent probes are used for spotted DNA arrays, since different fluorophores can be used to label different RNA populations. These can be simultaneously hybridized to the same array, allowing differential gene expression to be monitored directly. Typically, Cy3 is used to label one probe and Cy5 to label the other. Cy3 fluoresces bright red and Cy5 fluoresces bright green. If a particular RNA is present only in the Cy3-labeled probe, the corresponding feature on the array appears red. If another RNA is present only in the Cy5-labeled probe, that spot appears green. RNAs found in both probes would hybridize in equivalent amounts, and these features would appear yellow. Dual labeling is not used on GeneChips. Instead, hybridization is carried out with separate probes on two identical chips and the signal intensities are measured and compared by the accompanying analysis software. The process of measurement of differential gene expression levels using a microarray is illustrated in *Fig. 1*.

2D protein gels

2D-PAGE is a well-established biochemical technique in which proteins are separated on the basis of two separate properties: their **isoelectric point (pI)** and their **molecular mass**. Separation in the first dimension is nowadays carried out by isoelectric focusing in an immobilized pH gradient. The pH gradient is generated by a series of buffers, and an **immobilized pH gradient** is produced by covalently linking the buffering groups to the gel, therefore preventing migration of the buffer itself during electrophoresis. **Isoelectric focusing** means allowing proteins to migrate in an electric field until the pH of the buffer is the same as the pI of the protein. The pI of the protein is the pH at which it carries no net charge and therefore does not move in the applied electric field. Next, the gel is equilibrated in the detergent **sodium dodecylsulfate (SDS)**, which binds uniformly to all proteins and confers a net negative charge. Therefore, separation in the second dimension can be carried out on the basis of molecular mass.

After the second-dimension separation, the protein gel is stained with a universal dye to reveal the position of all protein spots (see the illustration in *Fig. 2*). Reproducible separations can then be carried out with similar samples to allow comparison of protein expression levels. Each protein can be character-

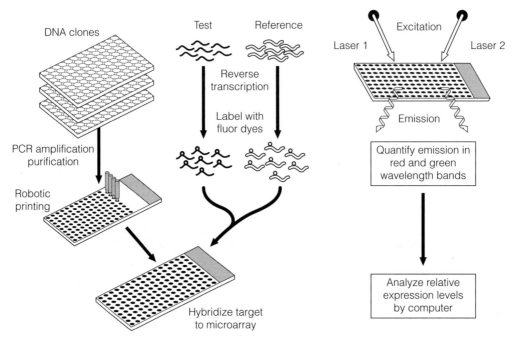

DNA clones

Test   Reference

Excitation

Laser 1   Laser 2

Reverse
transcription

Label with
fluor dyes

PCR amplification
purification

Robotic
printing

Emission

Quantify emission in
red and green
wavelength bands

Hybridize target
to microarray

Analyze relative
expression levels
by computer

Fig. 1. The process of differential expression measurement using a DNA microarray. DNA clones are first amplified and printed out to form a microarray. Test and reference RNA samples are then reverse transcribed and labeled with different fluor dyes (Cy5 and Cy3), which fluoresce in different (red/green) wavelength bands. These are hybridized to the microarray. Fluorescence of each dye is then measured for each feature (gene) using laser excitation, and converted to relative expression levels in the two samples. Reproduced with permission from Duggan D.J. et al., Expression profiling using cDNA microarrays, Nature Genet. **21***(Suppl. 1): pp. 10–14, 1999.*

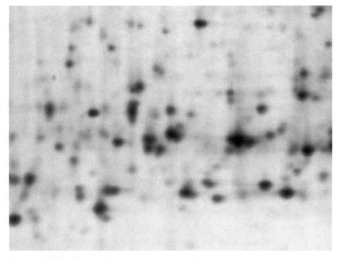

Fig. 2. A section from a 2D protein gel. The sample has been separated on the basis of isoelectric pH (horizontal dimension) and molecular mass (vertical dimension). Each spot should correspond to a single protein.

ized using x,y co-ordinates, where x corresponds to the pI and y to the molecular mass. Differential expression may be shown by the presence/absence of particular spots or by different spot intensities between samples. However, one complication that must be taken into account is that different forms of **post-translational protein modification** (e.g. phosphorylation, glycosylation) also contribute to position shifts and quantitative variations on 2D gels.

B4 PROTEIN INTERACTION DATA

Key Notes

The importance of protein interactions	Protein interactions underlie most cellular functions. Protein–protein interactions result in the formation of transient or stable multi-subunit complexes. An understanding of these complexes is required for the functional annotation of proteins and is a step towards the elucidation of molecular pathways such as signaling cascades and regulatory networks. Protein interactions with nucleic acids form an important area of study, since such interactions are required for replication, transcription, recombination, DNA repair and many other important processes. Proteins also interact with small molecules, which act as ligands, substrates, cofactors and allosteric regulators.
Genetic methods	Genetic analysis can be used to infer protein interactions. Suppressor mutants may compensate for a deleterious original mutation by restoring a disrupted protein interaction. Similarly, synthetic lethal effects may reflect the inability of two mutant proteins to interact. Dominant negative mutations show that a protein functions as a multimeric complex.
Affinity methods	Physical methods for the detection of protein interactions usually rely on the affinity of one protein for another. Affinity chromatography and co-immunoprecipitation each involve the use of a bait molecule to pull interacting proteins out of solution. These methods may identify indirect as well as direct interactions, but direct interactions can be proven by cross-linking.
Molecular and atomic methods	X-ray crystallography and nuclear magnetic resonance spectroscopy can help to characterize protein interactions at the atomic level. Other molecular methods for protein interaction analysis include FRET, SPR spectroscopy and SELDI. Many of these methods can be directly integrated with protein annotation by mass spectrometry.
Library-based methods	Library-based protein interaction assays have two major advantages: a highly parallel assay format and a direct link between candidate interacting proteins and their cDNAs. The predominant method is the Y2H system, in which interacting proteins are identified through their ability to assemble a functional transcription factor. Large amounts of data are generated in such experiments but there is a high level of false-positive and false-negative results, so further evidence is required to confirm candidate interactions.

Related topics:

Sequencing DNA, RNA and
   proteins (B1)
Protein structural data (B2)
Gene expression data (B3)
Microarray data: analysis
   methods (J1)

Analyzing protein mass
   spectrometry data (K2)
Molecular pathways (L1)
Protein interaction informatics (L2)

The importance of protein interactions

Most of the functions of a cell are carried out by proteins. These functions depend on **protein interactions**, that is, proteins interacting with each other and with other types of molecule. The disruption of protein interactions (by mutation, infection with a pathogenic organism, drug and chemical abuse, etc.) can have a variety of ill effects. Conversely, therapeutic drugs work by interacting beneficially with proteins.

It is useful to know about **protein–protein interactions** for several reasons. Many proteins either exist in permanent **multi-subunit complexes** (e.g. hemoglobin) or assemble into complexes in order to function (e.g. DNA polymerase, RNA polymerase). The functions of many other proteins are influenced by the formation of **transient complexes**. This includes many signaling proteins, and proteins involved in the regulation of processes such as the cell cycle and membrane transport. Through their interactions, proteins can influence each other's activity. This may be by binding (e.g. proliferating cell nuclear antigen [PCNA] stimulates the activity of DNA polymerase δ) or by catalysis (e.g. protein kinases dephosphorylate residues on other proteins and modulate their activities). The fact that the human genome appears to contain only 35 000–50 000 protein coding genes, compared with 4000 for the single *Escherichia coli* cell, indicates that organism complexity is achieved by means other than total gene number. The ability of proteins to interact in concert through diverse complexes and pathways is one way in which complexity can be achieved.

In addition to protein–protein interactions, it is also useful to know about the interactions between proteins and other molecules. **Protein–nucleic acid interactions**, for example, are central to a number of critical cellular processes (replication, recombination, transcription, RNA processing and protein synthesis, to name but a few). Interactions between proteins and small molecules are also important. This underlies many signaling processes (e.g. receptor–ligand binding) and often controls protein activity. For example, the activity of the *E. coli* TrpR repressor, which blocks transcription of genes involved in tryptophan biosynthesis, is regulated by the binding of tryptophan itself. Drug development (Topics N1 and N2) centers on the understanding and exploitation of protein–small ligand interactions.

In this topic, we briefly discuss some of the many different methods that can be used to characterize protein interactions. Bioinformatics is assuming an important role in this field as large amounts of interaction data accumulate. It is necessary to filter the data, extracting the reliable and meaningful interactions, build databases of interacting proteins, and use this information to reconstruct molecular pathways and networks. The use of interaction data is discussed in more detail in Topic L2.

Genetic methods

In genetically amenable species such as bacteria and yeast, protein–protein interactions can be inferred from **genetic analysis**. The basis of this type of approach is that interactions between two given proteins, X and Y, can be studied by looking at the behavior of mutations in their corresponding genes, X and Y. For example, a **suppressor mutation** restores the phenotype of a given primary mutation to normal or near normal. One explanation is that a primary mutation, in gene X, causes a conformational change in protein X that prevents its interaction with protein Y, therefore causing a loss of function. The suppressor mutation, in gene Y, introduces a complementary change in protein Y that restores the interaction and thus rescues the mutant phenotype (*Fig. 1a*). Another genetic test for protein interactions is the **synthetic lethal screen**. In this case, single mutations in genes X and Y in different individuals are nonlethal because the conformational changes in the corresponding proteins are minor and interaction between proteins X and Y is preserved. However, by combining the mutations in the same individual, the interaction between X and Y is disrupted and a lethal phenotype is observed (*Fig. 1b*).

The formation of a multi-subunit complex can be demonstrated using **dominant negative** mutants. In this case, a nonfunctional mutant form of the protein is produced in excess, for example by injecting mRNA into the cell or introducing a recombinant gene that is expressed at a high level. The mutant form of the protein sequesters the normal, active form into nonfunctional complexes resulting in a loss of function phenotype.

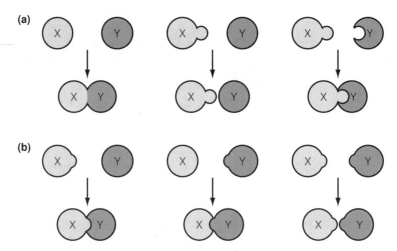

Fig.1. Genetic analysis of interactions between hypothetical proteins A and B. (a) A mutation that affects the structure of protein A and prevents interaction can be ameliorated by a suppressor mutation that introduces a compensatory change into protein B. (b) Mutations that affect the structure of proteins A and B can be tolerated individually, but result in a synthetic lethal effect when combined in the same cell.

Affinity methods   Genetic methods infer protein interactions but do not demonstrate them directly. In contrast, there are several physical methods that provide direct evidence for protein interactions by exploiting the **affinity** (tendency to bind specifically) of proteins for each other. Typically, one molecule is immobilized and used as a bait to trap interacting proteins, which can then be purified and characterized.

Affinity chromatography involves attachment of the bait molecule to a matrix, such as a Sepharose column. Protein extracts are then passed through the column and washed through with a low-salt buffer. Proteins that interact with the bait are retained, while others are eluted with the buffer. The bound proteins can themselves be eluted using a higher concentration of salt in the buffer. By using different salt concentrations, it is possible to discriminate between proteins that bind weakly and those that bind strongly to the bait. One popular variation of this method is known as **GST-pulldown**. In this technique, the bait protein is expressed as a fusion with the enzyme **glutathione-S-transferase**. This enzyme has a very high affinity for its substrate, **glutathione**, so the fusion protein will bind very strongly to glutathione-coated Sepharose beads. Proteins that interact with the GST-fusion protein will be retained on the chromatography column while noninteracting proteins are eluted. The advantage of this method is that any protein can be used as a bait, provided it can be expressed as a GST fusion. However, it is important to run a control chromatography experiment using GST alone, to weed out those proteins that interact with GST itself!

Another affinity method, **co-immunoprecipitation**, involves the precipitation of protein complexes using an antibody specific for one component of the complex. A cell lysate is prepared under conditions that preserve protein–protein interactions. An antibody raised against protein X is then mixed with the lysate, resulting in the precipitation of protein X and any proteins interacting with it. One problem with both affinity chromatography and co-immunoprecipitation is that these methods can isolate proteins that interact indirectly as well as those that interact directly with the bait. **Cross-linking** can be used to identify direct interactions and to probe the architecture of protein complexes.

Recent technological advances, led by the company Ciphergen, have allowed affinity assay formats to be extensively miniaturized, culminating in the development of **protein chips** (also called **protein microarrays** because they are similar in principle to the DNA microarrays discussed in Topic B3). Protein chips are manufactured in the same manner as DNA arrays, that is, by the robotic spotting of small amounts of liquid onto a solid substrate such as a glass slide. Capture agents may be broad in their scope (e.g. surface chemistries that capture whole classes of related proteins) or narrow and specific (e.g. antibodies).

Molecular and atomic methods

Protein–protein interactions and protein interactions with other molecules can be characterized using a number of molecular and atomic methods. For example, **X-ray crystallography** and **nuclear magnetic resonance spectroscopy** (Topic B2) can be used to determine the structure of proteins either as uncomplexed molecules or in complexes with ligands, cofactors and substrates. Another useful technique is **fluorescent resonance energy transfer (FRET)**, in which interaction between protein X and protein Y is shown by the transfer of energy from an excited donor fluorophore (conjugated to one protein) to an acceptor fluorophore (conjugated to the other). FRET occurs only when the two fluorophores are up to 10 nm apart, and can be detected by a change in the emission wavelength of the acceptor fluorophore. Instead of conjugated fluorophores, the interacting proteins may be fused to naturally occurring bioluminescent proteins (such as **green fluorescent protein, GFP**) in which case the technique may be termed **bioluminescent resonance energy transfer (BRET)**. The advantage of FRET/BRET is that it detects transient as well as stable interactions. **Mass spectrometry** (MS) can also be used to study protein–protein interactions. This type of experiment is usually carried out to determine the composition of a multi-subunit protein complex, and has been used for example to characterize the yeast spliceosome and anaphase-promoting complex. The procedure is to isolate an intact complex and subject it to matrix-assisted laser desorption/ionization (MALDI) MS to annotate individual proteins (Topic B1). This can be done either with or without prior electrophoretic separation of the proteins.

A final group of molecular methods is worth mentioning because the technologies are exploited on protein chips and can be integrated directly with MS analysis (Topic B1, Topic K2). One such method is **surface plasmon resonance (SPR) spectroscopy**, which is used on chips marketed by the US biotechnology company BIAcore. **Surface plasmon resonance** is an optical resonance phenomenon occurring when **plasmon waves** (charge density oscillations) become excited at the interface between a metal surface and a liquid. Interaction between protein X, which is immobilized on the metal surface, and protein Y, which is free in solution, is revealed by a change in the refractive index of the surface layer. Another method is **surface-enhanced laser desorption/ionization (SELDI)**. In this method, interacting proteins are trapped by affinity to a capture agent, such as an antibody, which is immobilized on a solid substrate. After washing and optional tryptic digestion of the interacting proteins, the substrate is coated with matrix compound and used directly for MALDI MS analysis (Topic B1, Topic K2). The advantage of SELDI over standard MALDI MS is that it provides more uniform mass spectra and allows protein quantification.

Library-based methods

All of the above methods are very powerful for detecting and characterizing protein interactions. However, their universal disadvantage is that only a few samples can be processed at any one time and the proteins identified as interactors cannot conveniently be associated with the genes that encode them. **Library methods** are based on traditional DNA expression libraries and solve both these problems by using a highly parallel assay format in which proteins are expressed from their associated cDNAs. Standard **expression libraries** and **phage display libraries** can be used for interaction analysis but in both cases, the interactions occur *in vitro*, where the protein folding and recognition conditions do not reflect those in the cell. Undoubtedly the method which has had the greatest impact on protein–protein interaction research is the **yeast two-hybrid (Y2H) system** and related technologies, in which interacting proteins are used to assemble a functional transcription factor. The basic principle of the Y2H system is shown in *Fig. 2*. Protein X is expressed as a fusion (a hybrid) with the DNA-binding domain of a transcription factor, to generate a **bait**. An expression

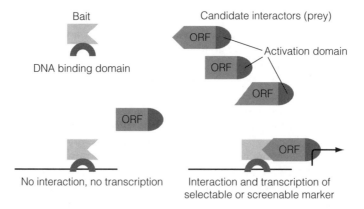

Fig. 2. Principle of the yeast two-hybrid system. (a) A bait protein is expressed as a fusion with the DNA-binding domain of a transcription factor, such as LexA. (b) A library of candidate interactors (prey) is expressed as fusions with a transcription factor's transactivation domain. (c) If there is no interaction between bait and prey, the reporter gene remains silent. (d) If the bait and prey interact, the reporter gene is also activated allowing the sequence of the interacting protein to be deduced from the cDNA.

library is then generated in which each clone is expressed as a fusion protein with the transcription factor's transactivation domain; these hybrids are potential **prey**. The final component of the system is a reporter gene that is activated specifically by the two-hybrid transcription factor. Mating between haploid yeast cells carrying the bait construct and those carrying the library of prey results in diploid cells carrying both components. In those cells where the bait interacts with the prey (protein Y) the transcription factor is assembled and the reporter gene activated, allowing the cells to be isolated and the DNA sequence of the interactor identified.

Y2H screens have been used to search for candidate interactions in a variety of molecular systems, including the entire yeast genome, several virus genomes and in specific sets of proteins from the fruit fly and the nematode worm. Very large amounts of data are generated in such screens. The main problem with this type of assay, and the reason that bioinformatics is so important, is the relatively high level of false-positive and false-negative results, meaning that accessory information has to be included to determine which interactions are genuine and which are spurious. The role of bioinformatics in protein interaction analysis is discussed in Topic L2.

C1 FILE FORMATS

Key Notes

Three common sequence formats	There are several conventions for representing nucleic acid and protein sequences, of which the NBRF/PIR, FASTA and GDE formats are widely used. These formats have limited facilities for comments, which must include a unique identifier code and the sequence accession number.
Files for aligned sequences	Aligned sequences can be represented in NBRF/PIR, FASTA or GDE formats but there are other formats devised especially for multiple sequence alignment, including MSF, PHYLIP and ALN.
Files of structural data	Structural data are maintained as flat files using the PDB format. Such files contain orthogonal atomic co-ordinates together with annotations, comments and experimental details.
Related topics	Annotated sequence databases (C2) Obtaining, viewing and analyzing Multiple sequence alignment and structural data (I4) family relationships (F1)

Three common sequence formats

If biological data is to be used by computer programs, it must be presented in a standard format that can be read by computer. It is very common to put data in text files. As the name suggests, these files contain text that can be read by a human being as well as a computer. They are rather like the files used by word-processing packages to hold documents, but there is one important difference: text files hold (almost) only the text and little auxiliary information about formatting (more details below). Here we discuss some standard formats and some more database specific formats are discussed in Topic C2.

Many bioinformatic databases and software applications are designed to work with sequence data, and this requires a standard format for inputting nucleic acid and protein sequence information. Three of the most common **sequence formats** are NBRF/PIR (National Biomedical Research Foundation/ Protein Information Resource), FASTA and GDE. Each of these formats has facilities not only for representing the sequence itself, but also for inserting a unique **code** to identify the sequence and for making **comments** which may include for example the name of the sequence, the species from which it was derived, and an accession number for GenBank or another appropriate database (Topic C2). *Figure 1* shows the same protein sequence, that of a guinea-pig serotonin receptor, represented in the three sequence formats listed above.

Figure 1a shows the **NBRF/PIR format**. Note that the first line begins with '>P1;' which specifies a protein sequence. If this was a nucleic acid sequence, it would begin with '>N1;'. The semicolon is followed by a code, in this case '5H1B_CAVPO', which is a unique sequence identifier. Serotonin is also known as 5-hydroxytryptamine, thus 5H1B identifies the protein as serotonin receptor 1B, while CAVPO identifies its source as the guinea-pig (*Cavia porcellus*). There

(a)
```
>P1;5H1B_CAVPO
Guinea pig serotonin receptor accession: O08892
MGNPEASCTP PAVLGSQTGL PHANVSAPPN NCSAPSHIYQ DSIALPWKVL LVVLLALITL
ATTLSNAFVI ATVYRTRKLH TPANYLIASL AFTDLLVSIL VMPISTMYTV TGRWTLGQAL
CDFWLSSDIT CCTASIMHLC VIALDRYWAI TDAVGYSAKR TPRRAAGMIA LVWVFSICIS
LPPFFWRQAK AEEEVLDCLV NTDHVLYTVY STGGAFYLPT LLLIALYGRI YVEARSRILK
QTPNKTGKRL TRAQLITDSP GSTSSVTSIN SRAPEVPCDS GSPVYVNQVK VRVSDALLEK
KKLMAARERK ATKTLGVILG AFIVCWLPFF IISLVMPICK DACWFHMAIF DFFTWLGYLN
SLINPIIYTM SNEDFKQAFH KLIRFKCTT
*
```

(b)
```
> 5H1B_CAVPO  O08892|guinea pig serotonin receptor
MGNPEASCTP PAVLGSQTGL PHANVSAPPN NCSAPSHIYQ DSIALPWKVL LVVLLALITL
ATTLSNAFVI ATVYRTRKLH TPANYLIASL AFTDLLVSIL VMPISTMYTV TGRWTLGQAL
CDFWLSSDIT CCTASIMHLC VIALDRYWAI TDAVGYSAKR TPRRAAGMIA LVWVFSICIS
LPPFFWRQAK AEEEVLDCLV NTDHVLYTVY STGGAFYLPT LLLIALYGRI YVEARSRILK
QTPNKTGKRL TRAQLITDSP GSTSSVTSIN SRAPEVPCDS GSPVYVNQVK VRVSDALLEK
KKLMAARERK ATKTLGVILG AFIVCWLPFF IISLVMPICK DACWFHMAIF DFFTWLGYLN
SLINPIIYTM SNEDFKQAFH KLIRFKCTT
```

(c)
```
%5H1B_CAVPO  O08892|guinea pig serotonin receptor
MGNPEASCTP PAVLGSQTGL PHANVSAPPN NCSAPSHIYQ DSIALPWKVL LVVLLALITL
ATTLSNAFVI ATVYRTRKLH TPANYLIASL AFTDLLVSIL VMPISTMYTV TGRWTLGQAL
CDFWLSSDIT CCTASIMHLC VIALDRYWAI TDAVGYSAKR TPRRAAGMIA LVWVFSICIS
LPPFFWRQAK AEEEVLDCLV NTDHVLYTVY STGGAFYLPT LLLIALYGRI YVEARSRILK
QTPNKTGKRL TRAQLITDSP GSTSSVTSIN SRAPEVPCDS GSPVYVNQVK VRVSDALLEK
KKLMAARERK ATKTLGVILG AFIVCWLPFF IISLVMPICK DACWFHMAIF DFFTWLGYLN
SLINPIIYTM SNEDFKQAFH KLIRFKCTT
```

Fig. 1. The sequence of a guinea-pig serotonin receptor in (a) NBRF/PIR format; (b) FASTA format; and (c) GDE format.

follows a **comment line**, and the rules allow this line to be of more or less any length so it can either be empty or far too wide to fit on a printed page. Then the sequence itself follows and is terminated by an asterisk (*). It is conventional to give files in this format the extension '.pir' or '.seq'.

Figure 1b shows the **FASTA format**. The first line begins with '>' but there is no designation of protein or nucleic acid sequence. The code is entered next and this is followed (on the same line) by comments, although it is conventional to delimit the comments with a '|' symbol. As with the NBRF/PIR format there is no limit to the length of the first line. One point to note about FASTA files is that they allow lower-case letters for the amino acids. Files in this format commonly have the extension '.fasta'.

Figure 1c shows the **GDE format**. This is essentially the same as the FASTA format, but the '>' symbol in the first line is replaced by '%'. Files in this format have the extension '.gde'.

All three file formats *ignore spaces and carriage returns*. This allows sequences to be typed out in a manner that is convenient for the user. In *Fig. 1*, for example, a space has been inserted every 10 amino acid residues and a carriage return after every 60, making it much easier to manually count the residues and identify amino acids at specific positions in the sequence. Note, however, that most standard word-processing software packages do not ignore blank spaces. For some purposes, it may be desirable or necessary to construct files from unpublished and preliminary sequence data, and if programs such as Microsoft Word or Corel WordPerfect are used, the results can be unpredictable. If using Word, use text only mode with a non-proportional font or, preferably, use a simple text editor such as Notepad.

To illustrate this point, consider the creation of the following very simple NBRF/PIR file:

```
>P1;MY_CODE
my cat
MYCATSATINMYLAP*
```

Despite the fact that this protein is clearly fictitious, the format is perfectly correct and it should be possible to search for the peptide sequence in other proteins. However, by typing this sequence into Microsoft Word and saving it as a Word document (cat.doc), the file proves to be over 19 thousand bytes in length, and therefore obviously contains much more than the simple text. By saving the file as *text with line breaks* (cat.txt), the file size is reduced to 39 bytes, which seems more reasonable. However, inspection of the contents of cat.txt reveals two extra characters at the end of each line and another at the very end of the file. It is therefore best to avoid word processors and use text editors for the preparation of sequences. If a word processor is used, the file should be saved as text and sent by ftp as ASCII (Topic O4), and a text editor should then be used on the computer where the sequence analysis is carried out, to check the integrity of the file. Another point to bear in mind is that the first line of a FASTA file and the second line of an NBRF/PIR file might be extremely long and it is essential not to cut it up by inserting carriage returns, otherwise the comments might be read as part of the sequence.

Files for aligned sequences

The output from **sequence-alignment programs** can be in any one of a number of formats. All three formats discussed above are suitable for dealing with aligned sequences but there are several formats designed specifically for alignment output. *Figure 2* shows partial results from the alignment of five serotonin receptor sequences, including the guinea-pig 5H1B receptor. In order to achieve the alignments, **gaps** must be introduced (Topic E3) and these are represented either by hyphens or dots. **Multiple sequence format (MSF)** is used by several software tools. **PHYLIP (phylogenetic inference package)** is the output format of the software of that name and CLUSTALW/X (F1) has its own **ALN format**. Multiple sequence alignment is discussed in more detail in Topic F1.

Files of structural data

The raw materials for bioinformatic studies on macromolecular structures are **PDB files**. These are text files using a format devised by the Protein Data Bank (Topic C4). Such files contain orthogonal atomic co-ordinates together with annotations, comments and experimental details. Examples of parts of such files are shown in *Fig. 3*. The most important aspect of PDB files is that the 'ATOM' lines are laid out in columns of characters *not* columns of words. Compare the first ATOM lines from *Fig. 3a, b* and *c* (only the left-hand parts of each line are displayed here):

```
ATOM    1   N   VAL     16   29.582    19.112    38.968
ATOM    1   N   ILE E   16   -9.947    23.613    20.817
ATOM    1   N   ALA      1   14.702   -10.824     3.425
```

The last three columns show the orthogonal co-ordinates (x, y, z) of the atom and they are deduced by counting the positions including the spaces, *not* by counting the words. This is because the x co-ordinate is the sixth word in the first and third cases but the seventh word in the second case, because a new

(a)
```
 MSF:    435   Type: P      Check:   2299    ..

 Name: 5H1A_MOUSE oo  Len:  435  Check:  7521  Weight:  0.166
 Name: 5H1A_RAT   oo  Len:  435  Check:  8470  Weight:  0.250
 Name: 5H1A_HUMAN oo  Len:  435  Check:  8517  Weight:  0.166
 Name: 5H1B_CAVPO oo  Len:  435  Check:   829  Weight:  0.222
 Name: 5H1B_CRIGR oo  Len:  435  Check:  6962  Weight:  0.100


5H1A_MOUSE      MD......MF  SLGQGNNTTT  SLEPFG....  ..TGGNDTGL  SNVTFSYQVI
5H1A_RAT        MD......VF  SFGQGNNTTA  SQEPFG....  ..TGGNVTSI  SDVTFSYQVI
5H1A_HUMAN      MD......VL  SPGQGNNTTS  PPAPFE....  ..TGGNTTGI  SDVTVSYQVI
5H1B_CAVPO      MGNPEASCTP  PAVLGSQTGL  PHANVSAPPN  NCSAPSHIYQ  DSIALPWKVL
5H1B_CRIGR      MEEQGIQCAP  PPPAASQTGV  PLVNLS...H  NCSAESHIYQ  DSIALPWKVL


5H1A_MOUSE      TSLLLGTLIF  CAVLGNACVV  AAIALERSLQ  NVANYLIGSL  AVTDLMVSVL
5H1A_RAT        TSLLLGTLIF  CAVLGNACVV  AAIALERSLQ  NVANYLIGSL  AVTDLMVSVL
5H1A_HUMAN      TSLLLGTLIF  CAVLGNACVV  AAIALERSLQ  NVANYLIGSL  AVTDLMVSVL
5H1B_CAVPO      LVVLLALITL  ATTLSNAFVI  ATVYRTRKLH  TPANYLIASL  AFTDLLVSIL
5H1B_CRIGR      LVALLALITL  ATTLSNAFVI  ATVYRTRKLH  TPANYLIASL  AVTDLLVSIL
```

{ rest of file omitted }

(b)
```
     5      435
5H1A_MOUSE MD------MF  SLGQGNNTTT  SLEPFG----  --TGGNDTGL  SNVTFSYQVI
5H1A_RAT   MD------VF  SFGQGNNTTA  SQEPFG----  --TGGNVTSI  SDVTFSYQVI
5H1A_HUMAN MD------VL  SPGQGNNTTS  PPAPFE----  --TGGNTTGI  SDVTVSYQVI
5H1B_CAVPO MGNPEASCTP  PAVLGSQTGL  PHANVSAPPN  NCSAPSHIYQ  DSIALPWKVL
5H1B_CRIGR MEEQGIQCAP  PPPAASQTGV  PLVNLS---H  NCSAESHIYQ  DSIALPWKVL


           TSLLLGTLIF  CAVLGNACVV  AAIALERSLQ  NVANYLIGSL  AVTDLMVSVL
           TSLLLGTLIF  CAVLGNACVV  AAIALERSLQ  NVANYLIGSL  AVTDLMVSVL
           TSLLLGTLIF  CAVLGNACVV  AAIALERSLQ  NVANYLIGSL  AVTDLMVSVL
           LVVLLALITL  ATTLSNAFVI  ATVYRTRKLH  TPANYLIASL  AFTDLLVSIL
           LVALLALITL  ATTLSNAFVI  ATVYRTRKLH  TPANYLIASL  AVTDLLVSIL
```

{ rest of file omitted }

(c)
```
>P1;5H1A_MOUSE

MD------MFSLGQGNNTTTSLEPFG------TGGNDTGLSNVTFSYQVITSLLLGTLIF
CAVLGNACVVAAIALERSLQNVANYLIGSLAVTDLMVSVLVLPMAALYQVLNKWTLGQVT
CDLFIALDVLCCTSSILHLCAIALDRYWAITDPIDYVNKRTPRRAAALISLTWLIGFLIS
IPPMLGWRAPEDRSNPNECTISKDHG-YTIYSTFGAFYIPLLLMLVLYGRIFRAARFRIR
KTVKKVEKKGAGTSFGTSSAPPPKKSLNGQPGSGDCRRSAENRAVGTPCANGAVRQGEDD
ATLEVIEVHRVGNSKGDLPLPSESGATSYVPACLERKNERTAEAKRKMALARERKTVKTL
GIIMGTFILCWLPFFIVALVLPFCESSCHMPELLGAIINWLGYSNSLLNPVIYAYFNKDF
QNAFKKIIKCKFCR-
*

>P1;5H1A_RAT

MD------VFSFGQGNNTTASQEPFG------TGGNVTSISDVTFSYQVITSLLLGTLIF
CAVLGNACVVAAIALERSLQNVANYLIGSLAVTDLMVSVLVLPMAALYQVLNKWTLGQVT
CDLFIALDVLCCTSSILHLCAIALDRYWAITDPIDYVNKRTPRRAAALISLTWLIGFLIS
IPPMLGWRTPEDRSDPDACTISKDHG-YTIYSTFGAFYIPLLLMLVLYGRIFRAARFRIR
KTVRKVEKKGAGTSLGTSSAPPPKKSLNGQPGSGDWRRCAENRAVGTPCTNGAVRQGDDE
ATLEVIEVHRVGNSKEHLPLPSESGSNSYAPACLERKNERNAEAKRKMALARERKTVKTL
GIIMGTFILCWLPFFIVALVLPFCESSCHMPALLGAIINWLGYSNSLLNPVIYAYFNKDF
QNAFKKIIKCKFCRR
*
```

{ rest of file omitted }

Fig. 2. Partial results from the alignment of five proteins with CLUSTALW (Topic F1). The formats shown are (a) MSF output; (b) PHYLIP output; and (c) NBRF/PIR output.

(a) Trypsin

```
HEADER     HYDROLASE (SERINE PROTEINASE)              13-APR-88   1SGT     1SGT   3
COMPND     TRYPSIN (/SGT$) (E.C.3.4.21.4)                                  1SGT   4
SOURCE     (STREPTOMYCES $GRISEUS, STRAIN K1)                              1SGT   5
AUTHOR     R.J.READ,M.N.G.JAMES                                            1SGT   6
REVDAT   1   16-JUL-88 1SGT    0                                           1SGT   7
JRNL          There follow the literature references
REMARK   1  There follow several remarks of which only 1 is shown here    1SGT  21
REMARK   2 RESOLUTION. 1.7 ANGSTROMS.                                      1SGT  72
              The sequence (only 2 lines shown) follows
SEQRES   1     223  VAL VAL GLY GLY THR ARG ALA ALA ALA GLN GLY GLU PHE PRO  1SGT  92
SEQRES   2     223  PHE MET VAL ARG LEU SER MET GLY CYS GLY GLY ALA LEU    1SGT  93
FTNOTE   1  There follow several footnotes                                1SGT 110
HET    CA    246      1     CALCIUM ++ ION                                1SGT 151
FORMUL   2   CA    CA1 ++                                                 1SGT 152
FORMUL   3   HOH    *192(H2 O1)  The last 3 lines describe hetero atoms   1SGT 153
              there follow several lines of secondary structure assignment
              of which only the first is shown here
HELIX    1   A ALA      56  CYS      58  5                                1SGT 154
              There follow 7 lines describing the orthogonal coordinate system
ATOM     1  N   VAL    16      29.582  19.112  38.968  1.00 12.94         1SGT 199
ATOM     2  CA  VAL    16      30.031  20.461  38.668  1.00 15.43         1SGT
            ....the bulk of the file
ATOM  1618  CD1 LEU   245       2.571  16.977  47.866  1.00 40.15         1SGT1816
ATOM  1619  CD2 LEU   245       4.758  18.112  48.337  1.00 44.30         1SGT1817
ATOM  1620  OXT LEU   245       1.660  16.559  52.387  1.00 59.60         1SGT1818
TER   1621      LEU   245                                                 1SGT1819
HETATM 1622 CA   CA   246      14.219  32.828  30.463  1.00 13.21         1SGT1820
HETATM 1623 O    HOH    1      22.919  19.524  42.538  1.00  8.79         1SGT1821
            ....the remaining water molecules up to ...
HETATM 1814 O    HOH  192      -3.192  30.325  46.346  0.68 57.70         1SGT2012
CONECT   72  70                                        941                1SGT2013
            ....the connectivity data
MASTER      79   41    1    5   14   15    1    6 1813    1   30   18      1SGT2043
END                                                                       1SGT2044
```

(b) Complex of a proteinase ("E") with a polypeptide inhibitor ("I")

```
HEADER     COMPLEX(SERINE PROTEINASE-INHIBITOR)     21-JAN-83   3SGB     3SGBE  1
COMPND     PROTEINASE B FROM STREPTOMYCES GRISEUS (/SGPB$)                3SGB   4
COMPND   2 (E.C. NUMBER NOT ASSIGNED) COMPLEX WITH THIRD DOMAIN OF THE    3SGB   5
COMPND   3 TURKEY OVOMUCOID INHIBITOR (/OMTKY3)                           3SGB   6
SOURCE     (STREPTOMYCES $GRISEUS, STRAIN K1) AND TURKEY (MELEAGRIS       3SGB   7
SOURCE   2 GALLOPAVO)                                                     3SGB   8
AUTHOR     R.J.READ,M.FUJINAGA,A.R.SIELECKI,M.N.G.JAMES                   3SGB   9
There follow many lines of remarks and details of lit. references
One such remark is important
REMARK   2 RESOLUTION. 1.8 ANGSTROMS.                                     3SGB  73
Start of sequence ...
SEQRES   1 E 185  ILE SER GLY GLY ASP ALA ILE TYR SER SER THR GLY ARG    3SGB  92
Start of ATOM entries for "chain E" ...
ATOM     1  N   ILE E  16      -9.947  23.613  20.817  1.00 16.42         3SGB 156
...end of "chain E" and start of "chain I"
ATOM  1310  OXT TYR E 242     -10.317  35.858  21.204  1.00 29.02         3SGB1465
TER   1311      TYR E 242                                                 3SGB1466
ATOM  1350  N   ASP I   7      25.100  14.110  33.198  1.00 41.61         3SGB1467
ATOM  1351  CA  ASP I   7      25.863  15.369  33.122  1.00 40.76         3SGB1468
... to the end of the file
```

(c) Intestinal fatty acid binding protein

```
HEADER     FATTY ACID-BINDING                        20-FEB-98   1A57
TITLE      THE THREE-DIMENSIONAL STRUCTURE OF A HELIX-LESS VARIANT OF
TITLE    2 INTESTINAL FATTY ACID BINDING PROTEIN, NMR, 20 STRUCTURES
COMPND     MOL_ID: 1;
COMPND   2 MOLECULE: INTESTINAL FATTY ACID-BINDING PROTEIN;
         Many lines of remarks including the authors but 3 are included:
EXPDTA     NMR, 20 STRUCTURES
AUTHOR     R.A.STEELE,D.A.EMMERT,J.KAO,M.E.HODSDON,C.FRIEDEN,
AUTHOR   2 D.P.CISTOLA
         Compare the following line with its counterparts in (a) and (b)
REMARK   2 RESOLUTION. NOT APPLICABLE.
         Start of the sequence

SEQRES   1     116  ALA PHE ASP GLY THR TRP LYS VAL ASP ARG ASN GLU ASN
         Start of the secondary structure assignment
SHEET    1   A 5 GLY      4  LYS      7  0
MODEL        1   start of the atomic coordinates for "model 1"
ATOM     1  N   ALA     1      14.702 -10.824   3.425  1.00  0.00         N
ATOM     2  CA  ALA     1      13.562 -10.618   2.552  1.00  0.00         C
ATOM     3  C   ALA     1      12.273 -10.914   3.355  1.00  0.00         C
         ... up to the end of that and start of "model 2"
ATOM  2132  QG  GLU   116      15.846   0.773   9.258  1.00  0.00         Q
TER   2133      GLU   116
ENDMDL
MODEL        2
ATOM  2134  N   ALA     1      14.997  -8.697   2.368  1.00  0.00         N
ATOM  2135  CA  ALA     1      14.960  -9.149   3.746  1.00  0.00         C
         ... up to the end (there is a total of 20 such models).
```

Fig. 3. Parts of three PDB files showing (a) and (b) X-ray crystallographic data; and (c) NMR data. Comments not in the original files but added here for clarity are shown in italic.

word, the chain identifier (E), has been inserted. Other points that emerge from *Fig. 3* are: the atoms are numbered consecutively; amino acids are represented by three letters; the ends of the lines may or may not contain line numbers; NMR files do not have a *resolution* REMARK; and NMR files typically contain several models corresponding to different conformations.

C2 ANNOTATED SEQUENCE DATABASES

Key Notes

Primary sequence databases

The three primary sequence databases are GenBank (NCBI), the Nucleotide Sequence Database (EMBL) and the DNA Databank of Japan (DDBJ). These are repositories for raw sequence data, but each entry is extensively annotated and has a features table to highlight the important properties of each sequence. The three databases exchange data on a daily basis.

Subsidiary sequence databases

Particular types of sequence data are stored in subsidiaries of the main sequence databases. For example, ESTs are stored in dbEST, a division of GenBank. There are also subsidiary databases for GSSs and unfinished genomic sequence data.

Submission of sequences

Sequences may be submitted to any of the three primary databases using the tools provided by the database curators. Such tools include WebIn and Bankit, which can be used over the Internet, and Sequin, a stand-alone application.

SWISS-PROT and TrEMBL

SWISS-PROT is a collection of confirmed protein sequences with annotations relating to structure, function and protein family assignments. The related database TrEMBL is a translation of all coding sequences in the primary nucleic acid databases. The entries in TrEMBL are less extensively annotated than those in SWISS-PROT, but are moved to SWISS-PROT when reliable annotations become available.

Database interrogation

All the databases discussed in this topic can be searched by sequence similarity. However, detailed text-based searches of the annotations are also possible using tools such as Entrez. The simplest way to cross-reference between the primary nucleotide sequence databases and SWISS-PROT is to search by accession number, as this provides an unambiguous identifier of genes and their products.

Related topics

Useful bioinformatics sites on the WWW (A3)
Sequencing DNA, RNA and proteins (B1)
File formats (C1)
Genome and organism-specific databases (C3)

Miscellaneous databases (C4)
Data retrieval with Entrez and DBGET/LinkDB (D1)
Data retrieval with SRS (Sequence Retrieval System) (D2)
Database searches: FASTA and BLAST (E3)

Primary sequence databases

The **primary sequence databases** are repositories for raw sequence data, and can be accessed freely over the World Wide Web (WWW). There are three such databases, comprising the **International Nucleotide Sequence Database Collaboration**. These are **GenBank**, maintained by the **National Center for Biotechnology Information (NCBI)**, the **Nucleotide Sequence Database** maintained by the **European Molecular Biology Laboratory (EMBL)**, and the **DNA Databank of Japan (DDBJ)**. New sequences can be deposited in any of the databases since they exchange data on a daily basis.

The databases contain not only sequences but also extensive annotations. As an example, *Fig. 1* shows part of a GenBank file, in this case for the human gene *BTEB*. Much of the introductory part of the file is self-explanatory, containing information such as the locus name, the accession number, the source species, literature references and the date of submission. An important section of the file is the **features table**, which describes interesting features of the sequence. Since

```
LOCUS           HUMBTEB       4859 bp     mRNA              PRI         07-FEB-1999
DEFINITION      Human mRNA for GC box binding protein, complete cds.
ACCESSION       D31716
VERSION         D31716.1  GI:505081
KEYWORDS        GC box binding protein; zinc finger.
SOURCE          Homo sapiens germline cDNA to mRNA, clone_lib:placenta.
  ORGANISM      Homo sapiens
                Eukaryota; Metazoa; Chordata; Craniata; Vertebrata; Euteleostomi;
                Mammalia; Eutheria; Primates; Catarrhini; Hominidae; Homo.
REFERENCE       1
                {.........}
REFERENCE       2   (bases 1 to 4859)
  AUTHORS       Ohe,N., Yamasaki,Y., Sogawa,K., Inazawa,J., Ariyama,T., Oshimura,M.
                and Fujii-Kuriyama,Y.
  TITLE         Chromosomal localization and cDNA sequence of human BTEB, a GC box
                binding protein
  JOURNAL       Somat. Cell Mol. Genet. 19 (5), 499-503 (1993)
  MEDLINE       94120483
COMMENT         Submitted (31-May-1994) to DDBJ by:
                Yoshiaki Fujii-Kuriyama
                {.........}
FEATURES             Location/Qualifiers
     source          1..4859
                     /organism="Homo sapiens"
                     /db_xref="taxon:9606"
                     /clone_lib="placenta"
                     /germline
     gene            1265..1999
                     /gene="BTEB"
     CDS             1265..1999
                     /gene="BTEB"
                     /note="three-times repeated zinc finger motif"
                     /codon_start=1
                     /product="GC box binding protein"
                     /protein_id="BAA06524.1"
                     /db_xref="GI:1060891"
                      translation="MSAAAYMDFVAAQCLVSISNRAAVPEHGVAPDAERLRLPEREVT
                     KEHGDPGDTWKDYCTLVTIAKSLLDLNKYRPIQTPSVCSDSLESPDEDMGSDSDVTTE
                     SGSSPSHSPEERQDPGSAPSPLSLLHPGVAAKGKHASEKRHKCPYSGCGKVYGKSSHL
                     KAHYRVHTGERPFPCTWPDCLKKFSRSDELTRHYRTHTGEKQFRCPLCEKRFMRSDHL
                     TKHARRHTEFHPSMIKRSKKALANAL"
BASE COUNT      1285 a    1111 c    1193 g    1270 t
ORIGIN          Chromosome 9, q13.
          1 cacgttgggt gacataatgg ggttttttta attatagatt cacactgcat ttattcatca
                {...........}
       4801 ttcaccattg tggaatgatg ccctggcttt aaggtttagc tccacatcat gcttctctt
//
```

Fig. 1. GenBank entry for the human gene BTEB. Some information has been deleted from the file for the sake of brevity and is indicated thus {........}

GenBank is a nucleic acid repository, the fact that there is a protein-coding region is a feature in the entry. Note that *BTEB* is a very simple gene that has no introns. If there were introns, the CDS (coding sequence) feature would be more complicated: the entry would be extended to indicate the base positions of the exons, delimited by commas. For example, if there was a second exon encoding a further 20 amino acids residues, the CDS feature would read as follows: 1265..1999,2100..2159.

Subsidiary sequence databases

The main sequence databases have a number of subsidiaries for the storage of particular types of sequence data. For example, **dbEST** is a division of GenBank which is used to store **expressed sequence tags (ESTs)**, and an example entry in dbEST is shown in *Fig. 2*. Other divisions of GenBank include **dbGSS**, which is used to store single-pass **genomic survey sequences (GSSs)**, **dbSTS**, which is used to store **sequence tagged sites (STSs**; unique genomic sequences that can be used as physical markers) and the **HTG (high-throughput genomic) division**, which is used to store unfinished genomic sequence data. These types of sequences are discussed in more detail in Topic B1.

Submission of sequences

The robustness of data submitted to the primary sequence databases is important in the context of bioinformatics software. Clearly, the integrity of the scientists who submit the data is not readily checked by computers but errors must be avoided in database consistency. It is essential that the data are submitted in a supported format and that the submission is carried out by means of software provided by the database curators. Examples are **WebIn** provided by EMBL (www.ebi.ac.uk/embl/Submission) and **BankIt** provided by the NCBI (http://www.ncbi.nlm.nih.gov/BankIt/), each of which can be used to submit sequences to the databases over the WWW. A powerful stand-alone software tool, **Sequin**, is provided by the NCBI and can be used on UNIX, PC/Windows and Macintosh systems for sequence submission for those with no WWW access (http://www.ncbi.nlm.nih.gov/Sequin/index.html).

SWISS-PROT and TrEMBL

SWISS-PROT and the related database **TrEMBL** (Translated EMBL) are repositories for annotated protein sequences. *Figure 3* shows the SWISS-PROT entry for the BTEB protein, corresponding to the GenBank entry in *Fig. 1*. The entry contains large numbers of annotations, including a **features table** before the sequence. Each line begins with two letters, many of which are self-explanatory, for example ID (identity), AC (accession number), DT (date), DE (description), GN (gene name), CC (comment). Continuation lines are indicated by the symbols -!- at the start of a section and indents thereafter (this is shown in the CC field in *Fig. 3*). Characteristic features of SWISS-PROT entries include the DR (reference), KW (key words) and FT (features) fields. It is the presence of these careful and extensive annotations that makes SWISS-PROT so popular with biochemists. For example, in *Fig. 3*, there is a fairly comprehensive description of the protein and its function but also (in the DT field) cross-references to the relevant entries in the secondary databases PROSITE, PRINTS and Pfam (Topic F2).

SWISS-PROT provides the most up-to-date and extensively annotated information on protein sequences and its quality reflects its active management by human curators. TrEMBL (translated EMBL) is another database in the same format. The entries in TrEMBL are derived from translation of all coding sequences in the EMBL Nucleotide Sequence Database that are not already in

```
LOCUS        T48601          355 bp    mRNA           EST        06-FEB-1995
DEFINITION   yb01a01.s1 Stratagene placenta (#937225) Homo sapiens cDNA clone
             IMAGE:69864 3' similar to similar to gb:S71043_rna1 IG ALPHA-2
             CHAIN C REGION (HUMAN), mRNA sequence.
ACCESSION    T48601
VERSION      T48601.1  GI:650461
KEYWORDS     EST.
SOURCE       human.
  ORGANISM   Homo sapiens
             Eukaryota; Metazoa; Chordata; Craniata; Vertebrata; Euteleostomi;
             Mammalia; Eutheria; Primates; Catarrhini; Hominidae; Homo.
REFERENCE    1  (bases 1 to 355)
  AUTHORS    {....}
TITLE        Generation and analysis of 280,000 human expressed sequence tags
JOURNAL      Genome Res. 6 (9), 807-828 (1996)
MEDLINE      97044478
COMMENT      Other_ESTs: yb01a01.r1
             Contact: Wilson RK
             Washington University School of Medicine
             4444 Forest Park Parkway, Box 8501, St. Louis, MO 63108
             Tel: 314 286 1800
             Fax: 314 286 1810
             Email: est@watson.wustl.edu
             High qality sequence stops: 277
             Source: IMAGE Consortium, LLNL
             This clone is available royalty-free through LLNL ; contact the
             IMAGE Consortium (info@image.llnl.gov) for further information.
             Seq primer: -21m13
             High quality sequence stop: 277.
FEATURES             Location/Qualifiers
     source          1..355
                     /organism="Homo sapiens"
                     /db_xref="GDB:490761"
                     /db_xref="taxon:9606"
                     /clone="IMAGE:69864"
                     /sex="male"
                     /clone_lib="Stratagene placenta (#937225)"
                     /lab_host="SOLR cells (kanamycin resistant)"
                     /note="Organ: placenta; Vector: pBluescript SK-; Site_1:
                     EcoRI; Site_2: XhoI; Cloned unidirectionally.  Primer:
                     Oligo dT. Caucasian. Average insert size: 1.2 kb; Uni-ZAP
                     XR Vector; ~5' adaptor sequence: 5' GAATTCGGCACGAG 3' ~3'
                     adaptor sequence: 5' CTCGAGTTTTTTTTTTTTTTTTTT 3'"
BASE COUNT       62 a     117 c     98 g     69 t      9 others
ORIGIN
        1 ggcggctcag tagcaggtgc cgtccacctc cgccatgaca acagacacat tgacatgggt
       61 gggtttacca ccaagcgtcc gatggtcttc tgtgtgaagg ccagccaggc gcctccatgg
      121 caccatgcag gagaaggnct ccccccttctt ccagtcctcg gctgccacgc gcagtatgct
      181 ggtcacacga aggtcgtggt gccctggctg gntcctncan ggatgcccaa gtcaggtact
      241 tntcgcgggg cagctcctgt gacccctgca gccagcgaac cagcacgtcc ttggggcttn
      301 aagcngcgct accaggcact tcaaccgttc nccagcttcg ttcagggcca ncttc
//
```

Fig. 2. GenBank (dbEST) entry for a human EST clone. Some information has been deleted from the file for the sake of brevity and is indicated thus {.........}

SWISS-PROT. As further data ensure the reliability of annotations, TrEMBL entries are moved to SWISS-PROT.

Database interrogation

Detailed queries of the text annotation in the databases discussed above can be carried out using tools like SRS and Entrez (Section D). However, a comparison of *Figs 1* and *3* shows how a user can cross-reference between these databases. Let us assume, for example, that searching GenBank with a new sequence

```
ID   BTE1_HUMAN     STANDARD;     PRT;    244 AA.
AC   Q13886; Q16196;
DT   15-DEC-1998 (Rel. 37, Created)
DT   15-DEC-1998 (Rel. 37, Last sequence update)
DT   20-AUG-2001 (Rel. 40, Last annotation update)
DE   TRANSCRIPTION FACTOR BTEB1 (BASIC TRANSCRIPTION ELEMENT BINDING
DE   PROTEIN 1) (GC BOX BINDING PROTEIN 1) (KRUEPPEL-LIKE FACTOR 9).
GN   BTEB1 OR BTEB OR KLF9.
OS   Homo sapiens (Human).
OC   Eukaryota; Metazoa; Chordata; Craniata; Vertebrata; Euteleostomi;
OC   Mammalia; Eutheria; Primates; Catarrhini; Hominidae; Homo.
OX   NCBI_TaxID=9606;
RN   [1]
RP   SEQUENCE FROM N.A.
RX   MEDLINE=94120483; PubMed=8291025;
RA   Ohe N., Yamasaki Y., Sogawa K., Inazawa J., Ariyama T., Oshimura M.,
RA   Fujii-Kuriyama Y.;
RT   "Chromosomal localization and cDNA sequence of human BTEB, a GC box
RT   binding protein.";
RL   Somat. Cell Mol. Genet. 19:499-503(1993).
RN   [2]
RP   SEQUENCE OF 1-31 FROM N.A.
RX   MEDLINE=94327649; PubMed=8051167;
RA   Imataka H., Nakayama K., Yasumoto K., Mizuno A., Fujii-Kuriyama Y.,
RA   Hayami M.;
RT   "Cell-specific translational control of transcription factor BTEB
RT   expression. The role of an upstream AUG in the 5'-untranslated
RT   region.";
RL   J. Biol. Chem. 269:20668-20673(1994).
CC   -!- FUNCTION: TRANSCRIPTION FACTOR THAT BINDS TO GC BOX PROMOTER
CC       ELEMENTS. SELECTIVELY ACTIVATES MRNA SYNTHESIS FROM GENES
CC       CONTAINING TANDEM REPEATS OF GC BOXES BUT REPRESSES GENES WITH
CC       A SINGLE GC BOX.
CC   -!- SUBCELLULAR LOCATION: NUCLEAR.
CC   -----------------------------------------------------------------------
CC   This SWISS-PROT entry is copyright. It is produced through a collaboration
CC   between  the Swiss Institute of Bioinformatics  and the  EMBL outstation -
CC   the European Bioinformatics Institute.  There are no  restrictions on  its
CC   use  by  non-profit  institutions as long  as its content  is  in  no  way
CC   modified and this statement is not removed.  Usage  by  and for commercial
CC   entities requires a license agreement (See http://www.isb-sib.ch/announce/
CC   or send an email to license@isb-sib.ch).
CC   -----------------------------------------------------------------------
DR   EMBL; D31716; BAA06524.1; -.
DR   EMBL; S72504; AAD14110.1; -.
DR   MIM; 602902; -.
DR   InterPro; IPR000822; Znf-C2H2.
DR   Pfam; PF00096; zf-C2H2; 3.
DR   PRINTS; PR00048; ZINCFINGER.
DR   SMART; SM00355; ZnF_C2H2; 3.
DR   PROSITE; PS00028; ZINC_FINGER_C2H2_1; 3.
DR   PROSITE; PS50157; ZINC_FINGER_C2H2_2; 3.
KW   Transcription regulation; DNA-binding; Nuclear protein; Repeat;
KW   Zinc-finger; Metal-binding.
FT   DOMAIN       84    116       ASP/GLU-RICH (ACIDIC).
FT   DOMAIN      143    225       ZINC FINGERS.
FT   ZN_FING     143    167       C2H2-TYPE.
FT   ZN_FING     173    197       C2H2-TYPE.
FT   ZN_FING     203    225       C2H2-TYPE.
SQ   SEQUENCE   244 AA;  27234 MW;  2D1B5A5BB9D42221 CRC64;
     MSAAAYMDFV AAQCLVSISN RAAVPEHGVA PDAERLRLPE REVTKEHGDP GDTWKDYCTL
     VTIAKSLLDL NKYRPIQTPS VCSDSLESPD EDMGSDSDVT TESGSSPSHS PEERQDPGSA
     PSPLSLLHPG VAAKGKHASE KRHKCPYSGC GKVYGKSSHL KAHYRVHTGE RPFPCTWPDC
     LKKFSRSDEL TRHYRTHTGE KQFRCPLCEK RFMRSDHLTK HARRHTEFHP SMIKRSKKAL
     ANAL
//
```

Fig. 3. SWISS-PROT entry for the human protein BTEB, equivalent to the GenBank entry shown in Fig. 1. Note the DR field provides the EMBL accession number, allowing database entries to be cross-referenced.

obtained in the laboratory identifies the gene in *Fig. 1* as particularly interesting for a research project. How do we find the corresponding entry in SWISS-PROT? Note that the IDs are different, and, although SWISS-PROT has alternative GN entries (gene names), they do not correspond to the name on the GenBank file. The way to find the correct SWISS-PROT file is to search the SWISS-PROT database for the accession number D31716. Although SWISS-PROT has its own accession number, D31716 can be found as a DR field entry. The SWISS-PROT site or one of its mirrors will successfully locate the entry with D31716 as the search string.

C3 GENOME AND ORGANISM-SPECIFIC DATABASES

Key Notes

Organism-specific resources	As well as general databases that serve the entire biology community, there are many organism-specific databases that provide information and resources for those researches working on particular species. The number of organism-specific databases is growing as more genome projects are initiated, and many can be accessed from general genomics gateway sites such as GOLD.
Database formats	There is no universally agreed format for genome databases and several viewers and browsers have been developed with graphical displays for genomic sequence analysis and annotation. One of the most versatile formats is ACeDB (originally designed for the nematode *Caenorhabditis elegans*), which has an object-orientated database architecture and is now used in many applications outside the field of genomic bioinformatics.
Finding organism-specific databases	Organism-specific data are widely distributed on the Internet. In order to find and interrogate databases on specific organisms, it is necessary to use a gateway site to access relevant databases and information resources. Worked examples are provided, using GOLD as the gateway and illustrated with Ebola virus, the bacterium *Escherichia coli*, the fruit fly *Drosophila melanogaster* and the human genome.
Related topics	File formats (C1) Annotated sequence databases (C2) Miscellaneous databases (C4) Database management (O4)

Organism-specific resources

The annotated sequence databases discussed in Topic C2 are general to all organisms, and contain data relevant to viruses, bacteria, microbial eukaryotes, animals and plants, as well as recombinant molecules produced in the laboratory. However, there are also many databases devoted to particular organisms, and their numbers are increasing as further genome projects are initiated. Typically, such databases contain not only sequence data but also information on gene expression, mutant phenotypes, genome maps, genome sequencing projects and relevant scientific literature, and provide links to resources for obtaining clones, mutants and for contacting researchers. A selected list of organism-specific databases is provided in *Table 1*, but this represents only a small fraction of the resources available. The user interested in an organism that is not listed in *Table 1* could try using a search engine (Topic A3) to find a useful resource, but there are also a number of excellent gateways available on the WWW, which provide information on multiple organism-specific resources and links to the relevant sites. A number of these gateways are listed in *Table 2*.

Table 1. *A small selection of organism-specific genomic databases available on the WWW.*

Organism	Database/resource	URL
Escherichia coli	EcoGene	http://bmb.med.miami.edu/EcoGene/EcoWeb/
	EcoCyc (Encyclopedia of *E. coli* genes and metabolism)	http://ecocyc.pangeasystems.com/ecocyc/ecocyc.html
	Colibri	http://genolist.pasteur.fr/Colibri/
Bacillus subtilis	SubtiList	http://genolist.pasteur.fr/SubtiList/
Saccharomyces cerevisiae	*Saccharomyces* Genome Database (SGD)	http://genome-www.stanford.edu/Saccharomyces/
Plasmodium falciparum	PlasmoDB	http://PlasmoDB.org
Arabidopsis thaliana	MIPS *Arabidopsis* thaliana Database (MAtDB)	http://mips.gsf.de/proj/thal/db
	The *Arabidopsis* information resource (TAIR)	http://www.arabidopsis.org/
Drosophila melanogaster	FlyBase	http://flybase.bio.indiana.edu/
Caenorhabditis elegans	A *C. elegans* DataBase (ACeDB)	http://www.acedb.org/
Mouse	Mouse Genome Database (MGD)	http://www.informatics.jax.org/
Human	OnLine Mendelian Inheritance in Man (OMIM)	http://www.ncbi.nlm.nih.gov/omim

These databases are actively curated by members of the research community working on the particular organism of interest and generally include links to organism-specific resources such as clone sets and mutant strains.

Table 2. *Useful gateway sites providing information and links to multiple, organism-specific and genomic resources.*

Gateway site	URL
NCBI Genomic Biology	http://www.ncbi.nlm.nih.gov/Genomes/index.html
GOLD (Genomes OnLine Database)	http://wit.integratedgenomics.com/GOLD/
Organism-specific genome databases	http://www.unl.edu/stc-95/ResTools/biotools/biotools10.html
TIGR Microbial Database	http://www.tigr.org/tdb/mdb/mdbcomplete.html
Bacterial genomes	http://genolist.pasteur.fr/
Yeast databases	http://genome-www.stanford.edu/Saccharomyces/yeast_info.html
EnsEMBL genome database project	http://www.ensembl.org/
MIPS (Munich Information Center for Protein Sequences)	http://mips.gsf.de

Database formats

Genomic databases need to facilitate the storage and analysis of large amounts of data, but must also have a user-friendly front-end graphical display to allow relevant data to be displayed and analyzed. A number of viewers and browsers have been developed for genomic sequence analysis and annotation (*Table 3*). These include Artemis, Apollo, EnsEMBL and GoldenPath. One of the most versatile database formats is **ACeDB**. This was originally designed for research on the nematode worm *Caenorhabditis elegans* (**A *C. elegans* DataBase**). ACeDB has an object orientated database architecture (Topic O4) rather than a simple collection of data and it has been used to handle data on other organisms, including the yeast *Scizosaccharomyces pombe* and humans. ACeDB has a graph-

Table 3. Database tools for displaying and annotating genomic sequence data.

Viewer format	URL for further information and tutorials
Artemis	http://www.sanger.ac.uk/Software/Artemis
ACeDB	http://www.acedb.org/Tutorial/brief-tutorial.shtml
Apollo	http://www.ensembl.org/apollo/
EnsEMBL	http://www.ensembl.org
NCBI map viewer	http://www.ncbi.nlm.nih.gov/
GoldenPath	http://genome.ucsc.edu/

ical user interface with displays and tools designed for genomic data. Other features of ACeDB include **AQL (ACeDB query language)**, interfaces with Perl and Java, WWW interfaces (of which AceBrowser is the current and supported version), WinAce (the Windows95/NT version of ACeDB), CITA (a CORBA interface to the database) and Acembly (a sequence assembly system). The URL is http://www.acedb.org/.

Finding organism-specific databases

This section provides a number of worked examples of how to find organism-specific databases, resources and information. A good starting point for this type of search is the Genomes OnLine Database (GOLD). Once the GOLD top page has been accessed (http://wit.integratedgenomics.com/GOLD/), the site can be searched for information on any organism, for example Ebola virus, the bacterium *Escherichia coli*, the fruit fly *Drosophila melanogaster*, and humans.

Ebola virus
Information on Ebola virus can be found by clicking on the *Viruses* hyperlink under 'Other Links' on the GOLD top page. This accesses the European Bioinformatics Institute page of completed viral genomes, which currently lists nearly 700 viruses whose genomes have been sequenced, together with the corresponding European Molecular Biology Laboratory Nucleotide Sequence Database file (Topic C2), the sequence in FASTA format (Topic C1) and the sequence retrieval system entry (Topic D2). The URL is http://www.ebi.ac.uk/cgi-bin/genomes/genomes.cgi?genomes=viruses.

Escherichia coli
Entering *Escherichia coli* as a search term after accessing the SEARCH GOLD query form pulls records on nine different bacterial strains. Resources are listed in a table with the following headings: *Organism, Tree, Information, Size/ORF-number, Data Search, Institution, Funding, Genome Database, Status* and *Publication*. Of these, the links provided under the headings *Information* and *Genome Database* are the most useful. There are more than 20 resources listed including some general ones [e.g. National Center for Biotechnology Information (NCBI), SWISS-PROT, TIGR (The Institute for Genomic Research) and some specific ones (e.g. EcoCyc, EcoGene, Colibri, which are also shown in *Table 1*). The *E. coli* resources can be accessed directly via hyperlinks, and many of the sites have lists of further resources. For example, the EcoCyc page, the Encyclopedia of *E. coli* genes and metabolism, contains a link to 'other information on *E. coli*' which is a page containing 13 further links to *E. coli* resources on the WWW. Similarly, the Colibri site has a link 'other sites related to *E. coli*' which lists over 20 additional resources.

Drosophila melanogaster

When *Drosophila melanogaster* is used as a search term, the *Genome Database* links provided by GOLD are as follows: NCBI, BDGP (Berkeley *Drosophila* Genome Project), FlyBase-UK, FlyBase-USA, EBI-Proteome, KEGG and IBM-Annotation. Under *Information*, a taxonomy resource and another comprehensive *Drosophila* site, the Interactive Fly, are also listed. FlyBase-UK links to http://www.edgp.ebi.ac.uk/ and provides a large number of *Drosophila* resources, including clone orders, annotated sequences, raw sequence data in FASTA format, sequence sets and access to sequence analysis tools.

Human

When *Homo sapiens* is used as a search term, links provided by GOLD are as follows: NCBI, ORNL, RIKEN, EnsEMBL, Proteome-EBI, Sanger Centre, HOWDY and IBM-Annotation. Clicking on NCBI links to the NCBI human genome resources page, which contains images of human chromosomes. The user can select from a drop-down menu of clones, genes, physical maps, genetic maps and variation and then click on any of the chromosomes to see the information available for the chromosome chosen under that heading. The NCBI server also provides access to UniGene, Online Mendelian Inheritance in Man (OMIM) and other miscellaneous databases (Topic C4).

C4 MISCELLANEOUS DATABASES

Key Notes

Database resources

There are many types of database available to researchers in the field of biology. These include primary sequence databases for the storage of raw experimental data, secondary databases that contain information on sequence patterns and motifs, and organism-specific databases tailored for researchers working on a particular species. Other miscellaneous databases are discussed in this topic.

Specialized sequence databases

A number of databases have been developed for the storage and analysis of particular types of sequence, for example rRNA and tRNA sequences, introns, promoters and other regulatory elements.

OMIM

OMIM is the Online Mendelian Inheritance in Man database, a powerful resource for the study of human genetics and human molecular biology. Each OMIM entry has a full text summary of information known about a particular gene or trait, with links to primary sequence databases and other human genetics resources.

Incyte and UniGene

Incyte is a commercial database belonging to the LifeSeq® Foundation, providing gene sequences and transcripts with expert annotation. It is designed specifically for drug discovery research. UniGene is an experimental facility for the clustering of GenBank sequences and related to EST data. Currently six vertebrate and five plant species are covered by UniGene.

Structural databases

The primary resource for protein structural data is the PDB, which contains data derived from X-ray crystallography and NMR studies. Another structural database, the MMDB can be accessed at the NCBI web site using Entrez.

Proteins and higher-order functions

Many databases have been set up to store information on particular types of proteins, such as receptors, signal transduction components and enzymes. The compilation of data on different types of proteins, their functions and interactions, makes it possible to deduce higher-order functional networks in the cell, such as biochemical pathways, signal transduction systems and regulatory hierarchies. An example of such a combined database is KEGG.

Literature databases

Literature databases store scientific articles and allow various fields (title, authors, keywords, abstract) to be searched using text strings. Among the most widely used literature resources on the Internet are MEDLINE and PubMED, which cover the scientific literature from the 1960s up to the present day.

Related topics

Annotated sequence databases (C2)	Data retrieval with SRS (Sequence Retrieval System) (D2)
Genome and organism-specific databases (C3)	Protein families and pattern databases (F2)
Data retrieval with Entrez and DBGET/LinkDB (D1)	

Database resources

Databases are essentially large storage devices for scientific and other data. They can be searched and cross-referenced either over the Internet or using downloaded versions on local computers or computer networks. Specific types of database are discussed in different topics throughout this book. For example, the three primary nucleic acid databases [GenBank, the European Molecular Biology Laboratory (EMBL) Nucleotide Sequence Database and the DNA Databank of Japan] are discussed in Topic C2. These are called **primary databases** because they store raw sequence data. Similarly, SWISS-PROT and TrEMBL are the major primary databases for the storage of protein sequences. There are also secondary databases of protein families and sequence patterns, such as PROSITE, PRINTS and BLOCKS (Topic F2). These are called **secondary databases** because the sequences they contain are not raw data, but have been derived from the data in the primary databases. There are also many organism-specific databases containing information, links and resources dedicated to particular species (Topic C3). In this topic, we discuss some of the remaining database resources available, which can be grouped under the description **miscellaneous databases**. Note that the journal *Nucleic Acids Research* devotes its first issue every year to articles describing new databases and updates to existing ones. These articles can be accessed online at the following URL: http://www.nar.oupjournals.org/

Specialized sequence databases

The primary sequence databases are unbiased as to the type of sequence data they contain. However, a number of more specialized databases have been developed with particular types of nucleic acid or protein sequence in mind. For example, there are databases specifically for rRNA (ribosomal RNA) and tRNA (transfer RNA) sequences, for example the database of 5S rRNA sequences (http://biobases.ibch.poznan.pl/5SData/) and the database of small subunit rRNA sequences (http://rrna.uia.ac.be/ssu/). Further examples include databases for promoter sequences and other transcriptional regulatory elements, databases for regulatory elements in the noncoding region of mRNAs (messenger RNAs) and InBase, a database of inteins, which are small peptides that are spliced out of some microbial proteins (http://www.neb.com/neb/inteins.html).

OMIM

OMIM (Online Mendelian Inheritance in Man) is a comprehensive database of human genes and genetic disorders maintained by the National Center for Biotechnology Information (NCBI) and can be accessed at the following URL: http://www.ncbi.nlm.nih.gov/omim or through Entrez (Topic D1). Each OMIM entry has a full text summary of a gene or genetically determined phenotype and has numerous links to other databases such as the primary sequence databases, SWISS-PROT, PubMed references, general and locus-specific mutation databases, gene nomenclature databases and mapviewer. OMIM is an excellent starting point to find information on human genetics. An example of an OMIM file (*Fig. 1*) refers to the same gene/protein represented by *Figs 1* and *3* in Topic C2.

Incyte and UniGene

Incyte is an example of a **commercial database**. Unlike the **public databases** discussed above, which can be accessed freely by anyone using the WWW, commercial databases require subscription, as they are often the result of a single company's research and investment. **Incyte** is an integrated database of DNA sequences, transcripts, extensive annotations, expression data and access

```
1: *602902
BASIC TRANSCRIPTION ELEMENT-BINDING PROTEIN 1; BTEB1

Alternative titles; symbols
BTEB

Gene map locus 9q13

TEXT
The GC box is a common regulatory DNA element of eukaryotic genes. The
promoter region of rat CYP1A1 (108330) contains a single GC box within a
basic transcriptional element (BTE) required for constitutive expression of
the gene. By screening a liver library for the ability to bind BTE, Imataka
et al. (1992) isolated rat cDNAs encoding Sp1 (189906) and a protein that
they designated BTEB(BTE binding protein). Sequence analysis revealed that,
like Sp1, BTEB contains 3 consecutive zinc finger motifs. In transient
transfection experiments both BTEB and Sp1 stimulated promoters with
repeated GC boxes. However, the CYP1A1 promoter with only 1 GC box was
activated by Sp1 and repressed by BTEB. Ohe et al. (1993) used a rat BTEB
cDNA to screen a human placenta library and isolated cDNAs encoding BTEB1.
The sequences of the predicted 244-amino acid rat and human proteins are 98%
identical. Imataka et al. (1992) and Ohe et al. (1993) found that the mRNAs
encoding BTEB and BTEB1 contain a GC-rich leader sequence in the 5-prime
untranslated region that has the potential to form stem-loop structures and
that may control translation.

By analysis of a somatic cell hybrid panel and by fluorescence in situ
hybridization, Ohe et al. (1993) mapped the BTEB1 gene to 9q13.


REFERENCES
1. Imataka, H.; Sogawa, K.; Yasumoto, K.; Kikuchi, Y.; Sasano, K.;
Kobayashi, A.; Hayami, M.; Fujii-Kuriyama, Y. : Two regulatory proteins that
bind to the basic transcription element (BTE), a GC box sequence in the
promoter region of the rat P-4501A1 gene. EMBO J. 11: 3663-3671, 1992.
PubMed ID : 1356762

2. Ohe, N.; Yamasaki, Y.; Sogawa, K.; Inazawa, J.; Ariyama, T.; Oshimura,
M.; Fujii-Kuriyama, Y. : Chromosomal localization and cDNA sequence of human
BTEB, a GC box binding protein. Somat. Cell Molec. Genet. 19: 499-503, 1993.
PubMed ID : 8291025


CREATION DATE
Rebekah S. Rasooly : 7/29/1998

EDIT HISTORY
alopez : 7/29/1998


Copyright (c) 2000 Johns Hopkins University
```

Fig. 1. The OMIM file for the human gene BTEB.

to cDNA (copy DNA) clones for experimental studies. It is the property of the LifeSeq® Foundation and subscription information can be found at the following URL: http://www.incyte.com.

UniGene is another resource for genome research. In the words of its developers: 'UniGene is an experimental system for automatically partitioning GenBank sequences into a non-redundant set of gene-oriented clusters. Each UniGene cluster contains sequences that represent a unique gene, as well as related information such as the tissue types in which the gene has been expressed and its map location.' UniGene incorporates about 10^5 expressed sequence tags (ESTs; Topic B1) and is used by experimenters to design probes and reagents for gene mapping and expression analysis. The organisms included in UniGene were chosen on the basis of the availability of large amounts of EST data and to give a reasonable coverage of the vertebrate and plant kingdoms with examples of closely and distantly related species. These include human, mouse, cow, rat, zebrafish, *Xenopus*, wheat, rice, barley, maize and *Arabidopsis*.

Structural databases

Structural databases store data on protein (and nucleic acid) structure. The primary resource for protein structure data is the **Protein Data Bank (PDB)** available at the following URL: http://www.pdb.org/. This is the single world-wide archive of structural data and is maintained by the **Research Collaboratory for Structural Bioinformatics (RCSB)**, at Rutgers University. The associated **Nucleic Acid Data Bank (NDB)** is also maintained there. Data from both X-ray crystallography and nuclear magnetic resonance (NMR) spectroscopy studies (Topic B2) can be deposited in the PDB using a web-based interface called the **AutoDep Input Tool (ADIT)**. The data are extensively checked and verified by human curators before acceptance. An equivalent European database is the **Macromolecular Structure Database (MSD)** maintained by the European Bioinformatics Institute. The RSCB and MSD databases contain the same data.

There are also other structural databases such as Entrez's **Molecular Modeling Database (MMDB)** which aims to provide information on sequence and structure neighbors, links between the scientific literature and 3D structures, and sequence and structure visualization.

Proteins and higher-order functions

One important aim of bioinformatics is to use biological data to understand the higher-level functions of the cell, that is, biochemical pathways, regulatory networks, signal transduction pathways, and how these influence cell and organism behavior. A number of databases have been established with this goal in mind. Several databases have been designed, for example, to provide information on the functional annotation of proteins. Such databases include **PIR (Protein Information Resource)**, which can be accessed at http://pir.georgetown.edu/. Another valuable resource is the **Kyoto Encyclopedia of Genes and Genomes (KEGG)**, which is the primary resource of the Japanese GenomeNet service. KEGG is available at the following URL: http://www.genome.ad.jp/kegg/. The main database integrates a number of subsidiaries including **PATHWAY** (which stores data on molecular pathways and complexes), **GENES** (which stores functional information about genes and their products) and **LIGAND** (which stores data about chemical compounds and reactions occurring in the cell). Together, these data can be used for functional annotation and the grouping of genes and proteins into common pathways, networks and hierarchies.

There are also many databases that provide information on specific aspects of protein function. For example, **DIP (the Database of Interacting Proteins)** and **BIND (Biomolecular Interaction Network Database)** provide functional annotations of proteins on the basis of their interactions with each other and with other ligands. There are also databases for particular types of protein, for example ReBase, a database of restriction endonucleases and their target sites;

TRANSFAC, a database of transcription factors; Sentra, a database of signal transduction proteins; and NUREBASE, a database of nuclear receptors.

Literature databases

A literature database contains the abstracts and, in some cases, the full text and figures of published scientific articles. Such databases can be searched using text strings to find words in the title, abstract, keywords or main text, or by author or author's institution. One of the earliest comprehensive online library resources was **Medline**, which has been incorporated into a large resource called **PubMed** maintained by the NCBI. They are integrated into the Entrez suite of databases described in Topic D1. Other such resources include the **Web of Science**, which requires institutional subscription, and **BioMedNet** (http://www.bmn.com), which provides access to thousands of review articles published in the popular *Trends* and *Current Opinion* journals. One of the best features of BioMedNet is that reviews can be downloaded as .pdf files and stored on the computer like a personal library.

D1 DATA RETRIEVAL WITH ENTREZ AND DBGET/ LINKDB

Key Notes

Access to distributed data	Biological data is widely distributed over the WWW. As an alternative to standard search engines, dedicated data retrieval tools such as Entrez, DBGET and SRS can be used to search multiple biological databases and retrieve relevant information.
Entrez	Entrez is a WWW-based data retrieval tool developed by the NCBI, which can be used to search for information in 11 integrated NCBI databases, including GenBank and its subsidiaries, OMIM and the literature database MEDLINE, through PubMED.
Getting started with NCBI and Entrez	Entrez is accessed via the NCBI homepage and is a simple, user-friendly system. Text search terms (words or Boolean phrases) can be used to search individual databases, and sequences can be used as queries with utilities such as BLAST. Hits are listed in order of relevance or similarity, with hits on the target database known as neighbors and hits on other databases known as links.
DBGET/LinkDB	DBGET is a data retrieval tool maintained by Kyoto University and the University of Tokyo. It covers more than 20 databases and is closely associated with KEGG. A related system, LinkDB, finds relationships between entries in the various databases covered by DBGET and others. DBGET has a simpler and more limited search format than Entrez.
Related topics	Bioinformatics and the Internet (A3) Annotated sequence databases (C2) Miscellaneous databases (C4) Data retrieval with SRS (sequence retrieval system) (D2) Database searches: FASTA and BLAST (E3)

Access to distributed data

A large amount of biological information is available over the World Wide Web (WWW; Topic A2), but the data are widely distributed and it is therefore necessary for scientists to have efficient mechanisms for **data retrieval**. One approach is to use standard **search engines** to find relevant web pages (Topic A3). However, it is sometimes difficult to find the desired information using this method, especially if the chosen search term has other connotations and pulls out many irrelevant sites. Alternatively, there are a number of dedicated **data retrieval tools** that can be used to access information for molecular biologists. The most widely used of these are **Entrez** and **DBGET** (discussed in this topic) and **SRS (sequence retrieval system**; discussed in Topic D2). Each of these tools allows text-based searching of a number of linked databases as well as sequence searching with utilities such as BLAST (Topic E3). They differ in the databases they cover and how the retrieved information is accessed and presented.

Entrez

Entrez is a WWW-based data retrieval system, developed by the National Center for Biotechnology Information (NCBI), which integrates information held in all NCBI databases. These databases include nucleotide sequences (from GenBank and its subsidiaries), protein sequences, macromolecular structures and whole genomes. Other resources linked to the NCBI can also be searched using Entrez. These include OnLine Mendelian Inheritance in Man (OMIM; Topic C4) and the literature database MEDLINE, through PubMed. Entrez can be accessed via the NCBI web site at the following URL: http://www.ncbi.nlm.nih.gov/Entrez/. In total, Entrez links to 11 databases, which are listed in *Table 1*.

Getting started with NCBI and Entrez

Entrez is the common front-end to all the databases maintained by the NCBI and is an extremely easy system to use. The Entrez main page, as with all NCBI pages, is undemanding in its browser requirements and downloads quickly. Part of the front page is illustrated in *Fig. 1*. The databases available for searching can be accessed by hyperlinks at the top of the page, or by using the drop-down menu as shown. Once a database has been selected, a **search term** is then entered in the space provided. The search term may be a single word or a Boolean phrase. Clicking on 'GO' initiates the search. Hits in the selected database are displayed (these are known as **neighbors**) and matching records in other Entrez databases are also shown (these are known as **links**). Hits are ordered by similarity based on **precomputed analysis** of sequences/structures or the literature.

For the newcomer, the following URL provides an overview of Entrez and a useful tutorial: http://www.ncbi.nlm.nih.gov/Database/index.html. This page

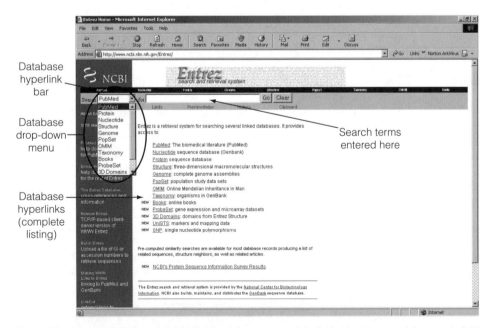

Fig. 1. The Entrez main page, showing the drop-down menu of available databases and the search field.

Table 1. The databases covered by Entrez, listed by category

Category	Database
Nucleic acid sequences	Entrez nucleotides: sequences obtained from GenBank, RefSeq and PDB
Protein sequences	Entrez protein: sequences obtained from SWISS-PROT, PIR, PRF, PDB, and translations from annotated coding regions in GenBank and RefSeq
3D structures	Entrez Molecular Modelling Database (MMDB)
Genomes	Complete genome assemblies from many sources
PopSet	From GenBank, set of DNA sequences that have been collected to analyse the evolutionary relatedness of a population
OMIM	OnLine Mendelian Inheritance in Man
Taxonomy	NCBI Taxonomy Database
Books	Bookshelf
ProbeSet	Gene Expression Omnibus (GEO)
3D domains	Domains from the Entrez Molecular Modelling Database (MMDB)
Literature	PubMED

is shown in *Fig. 2*, and includes a diagram showing the connectivity between eight of Entrez's databases.

DBGET/LinkDB   DBGET is an integrated data retrieval system developed and jointly maintained by the Institute for Chemical Research (Kyoto University) and the Human Genome Center (University of Tokyo). It is integrated with more than 20 data-

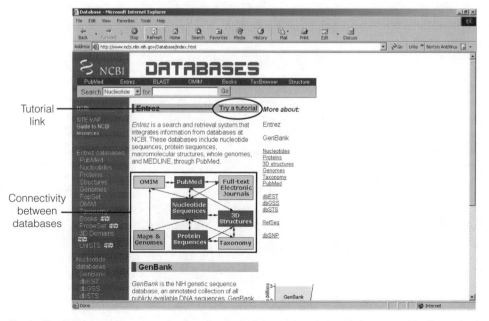

Fig. 2. The Entrez overview page, showing the tutorial link and the relationship between eight Entrez databases.

bases (*Table 2*), which can be searched one at a time or in combination using the commands **bfind** (for text searches) or **bget** (for searches based on accession number). Hits are presented as a list of results together with any available associated information. **LinkDB** is an associated database of links (**binary relationships**) between entries in the different databases available to DBGET and further organism-specific databases, such as the *C. elegans* Database (ACeDB), Flybase and the *Saccharomyces* Genome Database (SGD) (Topic C3). DBGET is closely associated with KEGG, the Kyoto Encyclopedia of Genes and Genomes, which is maintained by the same group (Topic C4).

Table 2. The databases covered by DBGET/LinkDB, listed by category

Category	Database
Nucleic acid sequences	GenBank, EMBL
Protein sequences	SWISS-PROT, PIR, PRF, PDBSTR
3D structures	PDB
Sequence motifs	PROSITE, EPD, TRANSFAC
Enzyme reactions	LIGAND
Metabolic pathways	PATHWAY
Amino acid mutations	PMD
Amino acid indices	AAindex
Genetic diseases	OMIM
Literature	LITDB Medline
Organism-specific gene catalogs	*E. coli, H. influenzae, M. genitalium, M. pneumoniae, M. jannaschii, Synechocystis, S. cerevisiae*

D2 DATA RETRIEVAL WITH SRS (SEQUENCE RETRIEVAL SYSTEM)

Key Notes

SRS and Entrez/DBGET	SRS (sequence retrieval system) is a data retrieval tool that, like Entrez and DBGET, can be used over the WWW. However, unlike these other systems, SRS is open source software and can be installed and run on a local computer network.
Using SRS	SRS databases are grouped but using different principles to those used by Entrez and DBGET. For example, all sequences (nucleic acid and protein) are grouped together, while these are separated by Entrez. The use of SRS involves selecting one or more of these groupings and, within each selected group, selecting one or more of the available databases. Queries can be submitted using two styles of query form, Standard or Extended.
Installing SRS	The advantage of SRS is that it can be installed locally. This allows SRS to be tuned to local databases, which use novel data formats. The tuning of SRS to deal with local databases involves programming in SRS's own scripting language, Icarus.
Related topics	Data retrieval with Entrez and DBGET/LinkDB (D1) Annotated sequence databases (C2) Genome and organism-specific databases (C3) Miscellaneous databases (C4) Protein families and pattern databases (F2)

SRS and Entrez/DBGET

SRS (sequence retrieval system) is a retrieval tool developed by the European Bioinformatics Institute (EBI) that integrates over 80 molecular biology databases. These are listed at http://srs6.ebi.ac.uk/srs6bin/cgi-bin/wgetz?-page+databanks+-newId, and are summarized in *Table 1*. Like Entrez and DBGET, SRS can be used over the WWW. However, the difference between SRS and Entrez/DBGET is that SRS is **open source software** that can be downloaded and installed locally. The result is that the databases and utilities that are available to the end user are not restricted by the activities of curators [at the National Center for Biotechnology Information (NCBI) in the case of Entrez, or GenomeNet in the case of DBGET]. Several large sites, including SWISS-PROT, use SRS as standard. The main ExPASy site (www.expasy.ch), one of the most useful bioinformatics gateways listed in Topic A3, is an example of SRS in action.

Using SRS

To start using SRS, access the appropriate site on your local network (if SRS has been installed locally) or alternatively try the SRS homepage at

Table 1. The databases covered by the SRS at http://srs6.ebi.ac.uk, listed by category

SRS description	Examples
Literature	MEDLINE, GO, GOA
Sequence	EMBL, EMBLNEW, SWISSPROT, SPTREMBL, REMTREMBL, TREMBLNEW, ENSEMBL, PATENT_PRT, USPO_PRT, IMGTLIGM, IMGTHLA
InterPro&Related	INTERPRO, IPRMATCHES, IPRMATCHES_ENSEMBL, PROSITE, PROSITEDOC, BLOCKS, PRINTS, PFAMA, PFAMB, PFAMHMM, PFAMSEED, PRODOM
SeqRelated	UTR, UTRSITE, TAXONOMY, GENETICCODE, EPD, HTG_QSCORE, CPGISLAND, EMBLALIGN, EMESTLIB
TransFac	TFSITE, TFFACTOR, TFCELL, TFCLASS, TFMATRIX, TFGENE
User Owned Databanks	USERDNA, USERPROTEIN
Application Results	FASTA, FASTX, FASTY, NFASTA, BLASTP, BLASTN, CLUSTALW, NCLUSTALW, PPSEARCH, RESTRICTIONMAP
Protein3DStruct	PDB, DSSP, HSSP, FSSP, PDBFINDER, RESID
Genome	MOUSE2HUMAN, LOCUSLINK, HGNC, HSAGENES
Mapping	RHDB, RHDBNEW, RHEXP, RHMAP, RHPANEL, OMIMMAP
Mutations	OMIMALLELE, MUTRES, SWISSCHANGE, EMBLCHANGE, MUTRESSTATUS, OMIM, OMIMOFFSET, HUMUT, HUMAN_MITBASE, P53LINK
Locus Specific Mutations	*41 entries omitted here*
SNP	MITSNP, dbSNPSubmitter, dbSNPAssay, dbSNPSNP, HGBASE, HGBASE_SUBMITTER
Metabolic Pathways	LENZYME, LCOMPOUND, PATHWAY, ENZYME, EMP, MPW, UPATHWAY, UREACTION, UENZYME, UCOMPOUND, UIMAGEMAP
Others	REBASE, SRSFAQ, BIOCATAL
System	PRISMASTATUS

http://srs6.ebi.ac.uk. On the WWW version, the top page lists 17 **classifications**. By clicking on the adjacent '+' symbol, each classification can be expanded to reveal the associated databases, and by clicking on the '–' symbol, the classifications can be collapsed. It is also possible to expand collapse all classifications. Note that SRS classifications, unlike those in Entrez, are grouped by the type of data not the type of molecule. Therefore, the *Sequence Libraries* classification covers all sequences (nucleotide and protein) whereas these data are separated by Entrez. Most of the other classifications are self-explanatory, but new users may find some unfamiliar. For example, *InterPro&Related* refers to the secondary databases of protein motifs (Topic F2), *SeqRelated* refers to specialized sequence databases such as UniGene (Topic C4) and *TransFac* refers to transcription factor databases (Topic C4).

To search with SRS, expand the relevant classifications and check boxes corresponding to the required databases as shown in *Fig. 1*. Clicking on the 'standard' or 'expanded' buttons then brings up the query form ('standard' is recommended for newcomers). Search terms can be entered in a number of fields by selecting from the drop-down menu (e.g. Accession number, Description, Keywords, Organism) or alternatively it is possible to search all fields simultaneously. Up to four different search terms can be used, linked by Boolean operators across multiple fields, so quite specific searches can be carried out. After clicking the 'submit query' button, the query results are displayed as

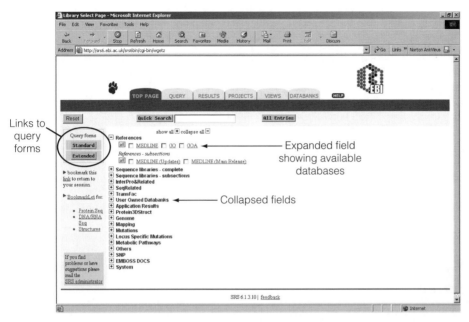

Links to query forms

Expanded field showing available databases

Collapsed fields

Fig. 1. The SRS top page at http://srs6.ebi.ac.uk. The user has selected the top page and has expanded the Literature classification. One or more of the databases (MEDLINE, GO and GOA) can now be selected and searched.

a table, with hyperlinks to files in the appropriate databases. A helpful tutorial for SRS users is available at http://srs6.ebi.ac.uk. From here, follow the "Information" link and look at the "User Guide".

Installing SRS

SRS is available to be installed locally. Essentially, it provides both the graphical interface and files to allow databases (such as those in *Fig. 1*) to be accessed. SRS works by constructing indexing files and should be updated regularly. The installation set comes with several configuration files and module files for the most popular bioinformatic databases ('databanks' in SRS). SRS configuration files establish which databases are available to SRS. Some of these files are used to add databases and others specify a grouping mechanism for databases or create new built-in views for data results. Module files correspond to the databases to be indexed.

Thus, an SRS system can be installed and operated with the minimum of effort. However, there are several points to consider when planning (or asking for help with) an SRS installation. Although the databases may all be remote, the indexing files are very large and, on a PC or other small system, the indexing process can take a lot of processor time (days). Nevertheless, a modest SRS system using example configuration module files is easily maintained and updated. The strength of SRS lies, in part, with providing the installer the capacity to use databases of his or her own even if they use a novel format. In order to do this, SRS has an associated scripting language called **Icarus**.

E1 SEQUENCE SIMILARITY SEARCHES

Key Notes

Sequence similarity searches

Sequence similarity searches of databases enable us to extract sequences that are similar to a query sequence. Information about these extracted sequences can be used to predict the structure or function of the query sequence. Prediction using similarity is a powerful and ubiquitous idea in bioinformatics. The underlying reason for this is molecular evolution.

Sequence alignment

Any pair of DNA sequences will show some degree of similarity. Sequence alignment is the first step in quantifying this in order to distinguish between chance similarity and real biological relationships. Alignments show the differences between sequences as changes (mutations), insertions or deletions (indels or gaps) and can be interpreted in evolutionary terms.

Alignment algorithms

Dynamic programming algorithms can calculate the best alignment of two sequences. Well-known variants are the Smith–Waterman algorithm (local alignments) and the Needleman–Wunsch algorithm (global alignments). Local alignments are useful when sequences are not related over their full lengths, for example proteins sharing only certain domains, or DNA sequences related only in exons.

Alignment scores and gap penalties

A simple alignment score measures the number or proportion of identically matching residues. Gap penalties are subtracted from such scores to ensure that alignment algorithms produce biologically sensible alignments without too many gaps. Gap penalties may be constant (independent of the length of the gap), proportional (proportional to the length of the gap) or affine (containing gap opening and gap extension contributions). Gap penalties can be varied according to the desired application.

Measurement of sequence similarity

Sequence similarity can be quantified using the score from the alignment algorithm, percentage sequence identities or more complex measures. The most useful statistical measures are discussed in Topic E2.

Similarity and homology

Similarity may exist between any sequences. Sequences are homologous only if they have evolved from a common ancestor. Homologous sequences often have similar biological functions (orthologs), but the mechanism of gene duplication allows homologous sequences to evolve different functions (paralogs).

Related topics

Amino acid substitution matrices (E2)
Database searches: FASTA and BLAST (E3)
Sequences filters (E4)

Herative database searches and PSI-BLAST (E5)
Multiple sequence alignment and family relationships (F1)

Sequence similarity searches

The sequences of biological macromolecules in modern databases are the products of molecular evolution. When sequences share a common ancestral sequence, they tend to exhibit similarity in their sequences, structures and biological functions. This is probably the most powerful idea in bioinformatics, because it enables us to make predictions. Often little is known about the function of a new sequence from a genome sequencing program, but if similar sequences can be found in the databases for which functional or structural information is available, then this can be used as the basis of a prediction of function or structure for the new sequence. It is very useful, therefore, to have search tools that take the new sequence as 'query', and search the database for similar sequences. Examples of such tools are the BLAST, FASTA and Smith–Waterman algorithms, which we will discuss later, but first we need to discuss how sequence similarity might be discovered, visualized and quantified.

Sequence alignment

Any pair of nucleic acid sequences will share a degree of similarity. For instance, DNA sequences are constructed from an alphabet of only four letters (A, T, G, C), so any sequence that consists of a mixture of these letters will show some similarity to any other similarly constructed sequence. We need to distinguish between this type of 'chance' similarity, and the similarity that is the result of a real evolutionary and/or functional relationship. We need a way of quantifying similarity.

Quantifying similarity begins with an **alignment**, such as that shown for the very short DNA sequences in *Fig. 1*. In this figure, the two sequences are written one above the other and the letters (defining DNA bases), which are vertically directly above and below each other are **aligned** or **equivalenced**. Reading the alignment from left to right, the first two As in each sequence are aligned, and these are followed by a T aligned with a C and then another pair of aligned Ts, and so on. It is important that not all letters in a particular sequence have equivalents in the other sequence. In *Fig. 1*, letters 7–9 of sequence 1 (TTG) are not aligned with any letters from sequence 2. When this happens, we say that a **gap** has been introduced. The point of introducing a gap here is that it enables a better alignment of the two sequences, in which more of the aligned pairs are identical letters. In this case, it enables the shared CGCAT directly after the gap to align between the two sequences.

Alignments can be interpreted in evolutionary terms. When identical letters are aligned, the simplest interpretation is that these letters were part of the ancestral sequence and have remained unchanged. When non-identical letters are aligned, the simplest interpretation is that a mutation has occurred in one of the sequences. Of course, in the absence of information about the ancestral sequence, it is not possible to know in which sequence the mutation actually occurred. Gaps in alignments can be interpreted in terms of the **insertion** or **deletion** of letters in one of the sequences with respect to the ancestral sequence. In *Fig. 1*, the gap could have resulted either from an insertion of three letters in sequence 1 or a deletion of three in sequence 2. Again, without the ancestral sequence, these possibilities cannot be distinguished, so the gap is sometimes referred to as an **indel**.

```
SEQ1:   AATTGATTGCGCATTTAAAGGG
SEQ2:   AACTGA---CGCATCTTAAGGG
```

Fig. 1. An alignment of two short DNA sequences.

Alignment algorithms

Finding the best alignment between the two sequences in *Fig. 1* was fairly straightforward, because the sequences are short and very closely related. It is much more difficult if the sequences are longer or if they are less closely related. In these cases, computational methods are required to find the best alignment of the sequences. Fortunately, there are known computational methods for this task, called **dynamic programming algorithms**. These algorithms take two input sequences and produce as output the best alignment between them.

Sequence similarity analyses commonly use two dynamic programming algorithms, the **Needleman–Wunsch algorithm** and the **Smith–Waterman algorithm**. These are closely related, but the main difference is that the Needleman–Wunsch algorithm finds **global** similarity between sequences, while the Smith–Waterman algorithm finds **local** similarity. An alignment from the Smith–Waterman algorithm might only cover a small (local) part of each sequence, while a Needleman–Wunsch alignment will try to cover as much of the sequences as possible, starting at left-most end of one of the sequences and finishing at the right-most end of one of the sequences. The Smith–Waterman algorithm is the most used because real biological sequences are often not similar over their entire lengths, but only in local portions. Examples of biological reasons for this are genes from different organisms with similar exons (local similarity) but different intron structures, or proteins that share only certain domains. Even though the Smith–Waterman algorithm is able to find these local similarities, it should also discover global sequence similarity if it exists.

Alignment scores and gap penalties

In their simplest incarnation, dynamic programming algorithms find alignments containing the largest possible number of identically matching letters, as in the alignment of *Fig. 1*. The problem, however, is that if it is permitted to insert gaps freely into the alignment then the algorithms produce alignments containing very large proportions of matching letters and large numbers of gaps. We need to control this, because insertion and deletion of monomers is a relatively slow evolutionary process, and alignments with large numbers of gaps do not make biological sense.

Dynamic programming algorithms get around this problem by using **gap penalties**. Essentially the algorithms form a score reflecting the quality of the alignment (a high positive score indicates a good alignment). A simple score contains a positive additive contribution of 1 for every matching pair of letters in the alignment, and a gap penalty is subtracted for each gap that has been introduced. In *Fig. 1*, if the gap penalty were equal to 1, then because there are 16 identically matching letters in the alignment and one gap, the score would be $16 - 1 = 15$. Dynamic programming algorithms find the alignment between the two sequences that maximizes the alignment score. This alignment depends on the choice of gap penalties: high gap penalties result in shorter, lower-scoring alignments with fewer gaps and lower penalties give higher-scoring, longer alignments with more gaps.

Several forms of gap penalty are commonly used. We described the simplest form above, where each gap attracted a **constant penalty** of 1, independent of the length of the gap. In general, a constant gap penalty can be written as A ($A=1$ above) where the size of A controls how strongly gaps are penalized. Sometimes a **proportional penalty** is used, where the penalty is proportional to the length of the gap. This penalty has the form Bl, where B is a constant (again controlling how strongly gaps are penalized), and l is the length of the gap. With this form, longer gaps attract larger penalties than shorter ones, which makes good biological sense. Finally, the most complex form of gap penalty is known as **affine**; this has both constant and proportional contributions, and it takes the form $A+Bl$. In this case, the constant A is called the **gap-opening penalty**, because it is applied to a gap of any length. The constant B is called the **gap-extension penalty**, because it is the penalty attached to extending the length of an existing gap by one unit. The motivation for the affine gap penalty is that opening a gap (penalty=A) should be strongly penalized, but once a gap is open it should cost less (B) to extend it.

It is not always clear what values should be used for the gap penalty constants A and B, and it does depend to some extent on the intended application. Most software for sequence alignment or similarity search comes with good default values for general sequence alignment or searching. In general, if you are interested in detecting only close sequence relationships then it would be a good idea to try increasing gap penalties above the default values to reduce the number of gaps. On the other hand, more distant relationships might be discovered by decreasing gap penalty values. It is often worth experimenting with different gap penalties around the default values. In some applications, gap penalties that differ significantly from default values should be used. One example would be the removal of vector sequence during genome sequencing (Topic B1): here the concern is only to find exact matches to the known vector sequence and very high gap penalties are appropriate.

Measuring sequence similarity

We have explained how dynamic programming algorithms attribute an alignment score that reflects the degree of relatedness of the aligned sequences. Some other measures are used, and perhaps the most common is the **percentage of identically aligned residues**. In *Fig. 1*, the number of identically aligned residues is 16, and the length of the alignment is 22, so the percentage sequence identity is $(16/22) \times 100 = 73\%$. This has the advantage that it is independent of the length of the alignment, so it can provide comparability between alignments of different length. However, it must always be remembered that high percentage identities are much more likely to reflect real biological or evolutionary relationships if they extend over long alignments. An alignment with 50% identity means little if its length is only 10 nucleotides, because it is likely to have occurred by chance. An alignment of 50% identity over 100 nucleotides is much more likely to be significant. Perhaps the most powerful methods of assessing sequence similarity are the statistical measures described in Topic E3.

Aligning sequences to maximize identical matches is reasonable for DNA sequences. However, with protein sequences, it is useful to take more account of the properties of the monomers that make up the sequence. Some amino acids substitute for each other better than others, and it is useful to take account of this in protein sequence alignment algorithms. This is discussed in Topic E2.

Similarity and homology

These terms are often confused. Any set of sequences can exhibit similarity and this may be quantified as discussed above. Sequences are **homologous** only if they evolved by divergence from a common ancestor. To say that two sequences share 50% homology is nonsense; to say that they share 50% similarity and that this indicates possible homology is the correct usage of the terms.

Sequence similarity searches are used very commonly to predict gene or protein function. The underlying theory is that similar sequences are likely to be homologous and therefore to have similar functions. If a biological function is essential for an organism to survive and reproduce, then loss of that function would be selected against by evolution, and this is the reason why homologous genes might be expected to retain the same function in different organisms. However, the evolutionary mechanism of gene duplication allows organisms to acquire redundant copies of genes. These redundant copies are then free to evolve new functions, and become homologous genes with different functions. This is the principle objection to prediction of gene function through sequence similarity, and is probably the reason for a significant amount of incorrect bioinformatics-derived annotation in the sequence databases. Examples of homologous genes with different functions are lysozyme (an enzyme) and α-lactalbumin (a mammalian regulatory protein). These proteins have very similar sequences, are almost certainly homologous and yet have very different functions.

When two homologous genes in different species have the same function, they are known as **orthologs**; when two genes in the same or different species have different functions they are known as **paralogs**.

E2 AMINO ACID SUBSTITUTION MATRICES

Key Notes

Maximizing amino acid identities

Protein sequences can be aligned to maximize amino acid identities, but this will not reveal distant evolutionary relationships.

Evolution

Protein-coding sequences evolve slowly compared with most other parts of the genome, because of the need to maintain protein structure and function. An exception to this is the fast evolution that might occur in the redundant copy of a recently duplicated gene.

Allowed changes

Changes in protein sequences during evolution tend to involve substitutions between amino acids with similar properties because these tend to maintain the structural stability of the protein.

Substitution score matrices

These matrices give scores for all possible amino acid substitutions during evolution. Higher scores indicate more likely substitutions. Example matrices are BLOSUM62 and PAM250. PAM stands for Accepted Point Mutations, and in this case, the evolutionary distance of the matrix is 250 amino acid changes per 100 residues. Dynamic programming algorithms for sequence alignment can operate using scores from these matrices.

Significance

Substitution score matrices allow detection of distant evolutionary relationships between protein sequences. It is possible to detect much more distant relationships by comparing protein sequences than by comparing nucleic acid sequences.

Visualization

Dot plots are a very good way to visualize sequence similarity and find repeats.

Related topics

Sequence similarity searches (E1)
Database searches: FASTA and
  BLAST (E3)

From *Instant Notes in Biochemistry*:
  Amino Acids (B1)

Maximizing amino acid identities

Protein sequences are constructed from an alphabet of 20 naturally occurring amino acids. Like nucleic acid sequences (see previous section), protein sequences can be aligned to maximize the number of identically matching pairs within the alignment. This is a good way of aligning closely related sequences. In this case, the contribution of every aligned pair of identical amino acids to the alignment score is one, and the contribution of every aligned pair of non-identical amino acids is zero. An example alignment is shown in *Fig. 1*.

Evolution

Evolution changes nucleic acid sequences, but protein-coding sequences evolve under strong functional constraints. Changes to these sequences generally survive only if they do not have a deleterious effect on the structure and function of the protein, because loss of the function of a protein is usually a disadvantage to the organism. An obvious exception to this occurs after a gene duplication event, when one copy of the gene can evolve very quickly, perhaps to become a pseudo-gene, or perhaps to gain a new and useful function. Usually, however, protein-coding genes evolve much more slowly than most other parts of any genome, because of the need to maintain protein structure and function.

Allowed changes When evolutionary changes do occur in protein sequences, they tend to involve substitutions between amino acids with similar properties, because such changes are less likely to affect the structure and function of the protein. For instance, hydrophobic amino acids of similar size tend to substitute for each other quite well, because, more often than not, these occupy positions within the hydrophobic core of the protein, where tight packing and hydrophobicity strongly affect the stability of the protein structure. Examples of such mutations would be changes between LEU and ILE or VAL. Protein sequences from within the same evolutionary family usually show substitutions between amino acids with similar physicochemical properties. *Fig. 2* shows physicochemical relationships between amino acids.

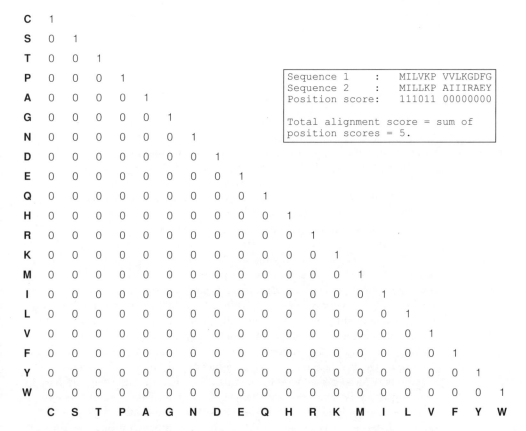

```
Sequence 1       :    MILVKP VVLKGDFG
Sequence 2       :    MILLKP AIIIRAEY
Position score:       111011 00000000

Total alignment score = sum of
position scores = 5.
```

	C	S	T	P	A	G	N	D	E	Q	H	R	K	M	I	L	V	F	Y	W
C	1																			
S	0	1																		
T	0	0	1																	
P	0	0	0	1																
A	0	0	0	0	1															
G	0	0	0	0	0	1														
N	0	0	0	0	0	0	1													
D	0	0	0	0	0	0	0	1												
E	0	0	0	0	0	0	0	0	1											
Q	0	0	0	0	0	0	0	0	0	1										
H	0	0	0	0	0	0	0	0	0	0	1									
R	0	0	0	0	0	0	0	0	0	0	0	1								
K	0	0	0	0	0	0	0	0	0	0	0	0	1							
M	0	0	0	0	0	0	0	0	0	0	0	0	0	1						
I	0	0	0	0	0	0	0	0	0	0	0	0	0	0	1					
L	0	0	0	0	0	0	0	0	0	0	0	0	0	0	0	1				
V	0	0	0	0	0	0	0	0	0	0	0	0	0	0	0	0	1			
F	0	0	0	0	0	0	0	0	0	0	0	0	0	0	0	0	0	1		
Y	0	0	0	0	0	0	0	0	0	0	0	0	0	0	0	0	0	0	1	
W	0	0	0	0	0	0	0	0	0	0	0	0	0	0	0	0	0	0	0	1

Fig. 1. Alignment to maximize amino acid identities, and the associated substitution score matrix. The matrix is able to align the first half of sequence 1 with the first half of sequence 2 because they are closely related with many identities. The second halves are poorly aligned because the relationship between the sequences is much weaker. There are no identities and the score matrix does not show how chemically reasonable matches could be made.

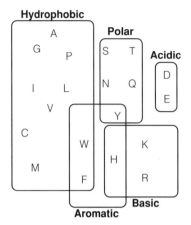

Fig. 2. Relationships between the physicochemical properties of amino acids.

Substitution score matrices

Substitution score matrices are used to show scores for amino acid substitutions. For example, *Fig. 1* shows a very simple score matrix, in which identical substitutions score 1 and nonidentical ones score zero. More general matrices show the likelihood of substitution occurring between each pair of amino acids. They enable the detection of similarity between protein sequences that would be missed by the simple identity matrix of *Fig. 1*.

An example is the **PAM250 substitution matrix** (*Fig. 3*): its elements are scores for alignment of (i.e. substitution between) each possible pair of amino acids. For instance, the score for aligning the similar amino acids L and I is high (2), reflecting the high likelihood of this substitution as an evolutionary process, while the score for aligning the dissimilar amino acids G and W is low (–7). Notice that the scores for alignments of identical amino acids vary. These reflect the frequency of occurrence of the amino acids in natural protein sequences. The alignment of two identical uncommon amino acids (e.g. aligning W with W) is more likely to reflect an evolutionarily significant alignment than the alignment of two identical common ones (e.g. S with S), which would be likely to happen by chance in two unrelated protein sequences. Alignments of uncommon identical amino acids are therefore given higher scores.

The dynamic programming sequence alignment algorithms described in Topic E1 are not limited to aligning identical letters. They can operate with any system of scoring alignments between monomers that can be expressed as a substitution matrix. The only difference is that instead of scoring 1 for every match and 0 for every mismatch, substitution scores are taken from the matrix. The total alignment score that the algorithms maximize is the sum of the substitution scores in the alignment, with the usual gap penalty contribution.

PAM stands for 'accepted point mutations'; the P and A are swapped to make it easier to say. PAM250 refers to an evolutionary distance of 250 accepted point mutations (PAMs) per 100 amino acid residues. PAM matrices are derived by counting observed evolutionary changes in closely related protein sequences, and then extrapolating the observed transition probabilities to longer evolutionary distances. It is possible to derive PAM matrices for any evolutionary distance, but in practice, the most commonly used matrices are PAM120 and PAM250. PAM matrices with smaller numbers represent shorter evolutionary distances. Choosing the matrix for the most appropriate evolutionary distance

C	12																		
S	0	2																	
T	-2	1	3																
P	-1	1	0	6															
A	-2	1	1	1	2														
G	-3	1	0	-1	1	5													
N	-4	1	0	-1	0	0	2												
D	-5	0	0	-1	0	1	2	4											
E	-5	0	0	-1	0	0	1	3	4										
Q	-5	-1	-1	0	0	-1	1	2	2	4									
H	-3	-1	-1	0	-1	-2	2	1	4	3	6								
R	-4	0	-1	0	-2	-3	0	-1	-1	1	2	6							
K	-5	0	0	-1	-1	-2	1	0	0	1	0	3	5						
M	-5	-2	-1	-2	-1	-3	-2	-3	-2	-1	-2	0	0	6					
I	-3	-1	0	-2	-1	-3	-2	-2	-2	-2	-2	-2	-2	2	5				
L	-6	-3	-2	-3	-2	-4	-3	-4	-3	-2	-2	-3	-3	4	2	6			
V	-2	-2	0	-1	0	-1	-2	-2	-2	-2	-2	-2	-2	2	4	2	4		
F	-4	-3	-3	-5	-4	-5	-4	-6	-5	-5	-2	-4	-5	0	1	2	-1	9	
Y	0	-3	-3	-5	-3	-5	-2	-4	-4	-4	0	-4	-4	-2	-1	-1	-2	7	10
W	-8	-2	-5	-6	-6	-7	4	7	7	5	3	2	-3	-4	-5	-2	-6	0	0	17
	C	S	T	P	A	G	N	D	E	Q	H	R	K	M	I	L	V	F	Y	W

```
Sequence 1      :    MILVKP -VVLKGDFG
Sequence 2      :    MILLKP AIIIRAEY-
Position score:      656256 044231370

Total alignment score = (sum of
position scores) - (gap penalty) =
54 - 1 = 53.
```

Fig. 3. The PAM250 matrix and the alignment of the sequences from Fig. 1 produced with this matrix. Total alignment scores for the two matrices should not be compared, but note that the PAM matrix is able to detect a much better alignment in the second halves of these sequences than the identity matrix of Fig. 1. With the introduction of a single gap, we see sensible alignments of hydrophobic amino acids, and alignments of K with R (both basic), D with E (both acidic) and F with Y (both aromatic).

might result in the best possible alignment of two sequences, but in practice, it is rarely possible to know what the evolutionary distance is, and experience shows that the PAM250 matrix usually produces reasonable alignments.

The **BLOSUM** series of matrices, and in particular BLOSUM50 and BLOSUM62, have recently become popular. These matrices have some theoretical advantages (their derivation does not involve the process of extrapolation to large evolutionary distances), and there is some evidence to suggest that they do outperform PAM matrices in practice. For this reason they are now more commonly used. The numbers attached to BLOSUM matrices do not have the same interpretation as those for PAM matrices (accepted mutations per 100 residues). In fact, BLOSUM matrices with smaller numbers represent longer evolutionary distances (BLOSUM50 represents a longer distance than BLOSUM62). There are several other matrices in fairly common use, and some specialized matrices (for instance PAM matrices specific to integral transmembrane proteins have been derivied).

Significance   A great deal of bioinformatics is concerned with the detection of evolutionary relationships between sequences. The use of matrices like BLOSUM62 extends our ability to detect distant relationships far beyond what could be found using the identity matrix in *Fig. 1*. This ability to detect very distant relationships (sometimes between sequences whose percentage of identical residues has diverged to below 30%) is the reason why sequence comparison should always be carried out at the protein rather than the nucleic acid sequence level, when this is possible. The ability to encode permissible changes in protein structures means that protein sequence alignment can reveal much more distant evolutionary relationships than naïve comparison of nucleic acid sequences.

Visualization

It is often useful to be able to visualize the similarity between two sequences. This can be done using dot plots, as illustrated in *Fig. 4*. In the case of *Fig. 4*, the dot plot has been used to discover internal similarity between *N*- and *C*-terminal portions of the same sequence, but it is equally possible to use dot plots to study similarity between two different sequences.

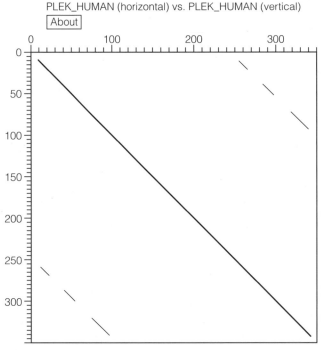

Fig. 4. A dot plot of the human pleckstrin sequence against itself produced with Erik Sonnhammer's 'dotter' program. The sequence is plotted from N- to C- terminus along the horizontal and vertical axes between residues 1 and approximately 350. Every residue pair with one amino acid from the horizontal sequence and one from the vertical sequence corresponds to a small square in the area of the plot. Such squares are colored according to the match score for the residue pair concerned. Darker colors correspond to better match scores and more similar residues. The dark line running diagonally from the origin to the lower right corner is made up of the squares corresponding to matches between residue 1 of the vertical sequence and residue 1 of the horizontal sequence, residue 2 and residue 2, and so on, and is dark colored because the horizontal and vertical sequences are identical in this case. More interesting are the faint lines running parallel to this in the top right and lower left corners of the plot. These indicate significant similarity between approximately residues 0–100 and 250–350 of the sequence and correspond to a repeated domain. In this case, this is the pleckstrin homology (PH) domain found in both N- and C-terminal parts of the pleckstrin sequence.

E3 DATABASE SEARCHES: FASTA AND BLAST

Key Notes

Database searches

Sequences similar to a query can be found in a database by aligning it to each database sequence in turn and returning the highest scoring (most similar) sequences. This can be achieved by dynamic programming algorithms but in practice faster approximate methods are often used.

Algorithms and software

BLAST and FASTA provide very fast searches of sequence databases. Unlike dynamic programming, they do not guarantee to find the best possible alignment to each database sequence, but in practice the effect on performance is usually minimal. Each operates by first locating short stretches of identically or near identically matching letters (words) that are eventually extended into longer alignments.

Statistical scores

The p value of a similarity score is the probability of obtaining a score at least as high in a chance similarity between two unrelated sequences of similar composition. Low p values indicate significant matches that are likely to have real biological significance. The related E value is the expected frequency of chance occurrences scoring at least as high as the identified similarity. A low p value for a similarity between two sequences can translate into a high E value for a search of a large database.

Sensitivity and specificity

These measures quantify the success of a database search strategy. Sensitivity measures the proportion of real biological sequence relationships in the database that were detected as hits in the search. Specificity is the proportion of the hits corresponding to real biological relationships. Changing E and p value thresholds results in a trade-off between these complementary measures of success.

Database types

Databases and query sequences can be protein or nucleic acid sequences and different query strategies are required for different types and combinations. In general, searches are more sensitive using strategies where protein-coding nucleic acid database and/or query sequences are first translated to protein sequences.

Worked example and program availability

In our opinion, the best BLAST server runs at the NCBI and can be used to search many general-purpose sequence databases. A similar FASTA implementation is available at the EBI. BLAST and FASTA are also used on many WWW sites to search organism-specific databases, for example at the Sanger Centre. We give a worked example using NCBI BLAST and FASTA at the EBI, which shows that a list of hits ranked by E value is useful, but that it is always necessary to inspect the associated sequence alignments to discover the region or domain of similarity between the sequences.

| **Related topics** | Annotated sequence databases (C2)
Sequence similarity searches (E1) | Amino acid substitution matrices (E2) |

Database searches

As detailed in Topic E1, sequence similarity searches aim to extract sequences from a database that are similar to a query sequence, perhaps with the aim of gathering family members, or predicting structure and function. In Topics E1 and E2, we discussed ways of quantifying the similarity between macromolecular sequences using sequence alignments and substitution score matrices. Similarity searches work by comparing (aligning) the query sequence to each database sequence in turn and then ranking the database sequences with the highest scoring (most similar) at the top. This process can be carried out by the dynamic programming algorithms described in Topic E1, but in practice, these algorithms are often too slow for searching large databases. We will discuss some alternative methods in this section. Also discussed in this section are the statistics of sequence similarity searching, and some issues surrounding database types and search parameters.

Algorithms and software

Dynamic programming algorithms are guaranteed to find the best alignment of two sequences for given substitution matrices and gap penalties. This is impressive, but the process is often quite slow, perhaps taking hours for a search of a large database. For this reason, alternative methods have been developed. Perhaps the most used of these are **FASTA** and **BLAST**; they are typically five (FASTA) to 50 (BLAST) times faster than dynamic programming. Unlike dynamic programming, these methods are not guaranteed to find the best alignment, and for this reason they could be less accurate. In practice, however, BLAST and FASTA usually suffer a minor degradation in accuracy when compared with dynamic programming, and the increased speed is often worth this small cost.

FASTA and BLAST are both based around similar ideas; in particular, they both make the reasonable assumption that high-scoring alignments are likely to contain short stretches of identical or near identical letters. These short stretches are called words. In the case of BLAST, the first step is to look for words of a certain fixed word length W that score higher than a certain threshold score (T). The value of W is normally 3 for protein sequences or 11 for nucleic acid sequences. W and T are under the control of the user, but there is rarely any need to change the default values. FASTA, on the other hand, looks for identically matching words of length *ktup*. *ktup* is under the control of the user, but the default values of 2 for protein sequences and 6 for DNA sequences are again seldom changed.

Both FASTA and BLAST employ a second step where extensions to the initially identified word matches are sought. BLAST extends individual word matches until the total score of the alignment falls from its maximum value by a certain amount producing an alignment without gaps. Early versions of BLAST always produced alignments without gaps, later versions now produce gapped alignments. The BLAST terminology for these high-scoring local alignments is **high-scoring segment pairs (HSPs)**.

FASTA tries to find ungapped alignments of the sequence that contain a high density of the initially identified word matches, and then attempts to join these into high-scoring gapped alignments. Finally, having identified high-scoring alignments between the sequences, FASTA produces its final alignment and score by a full dynamic programming alignment of the identified high-scoring regions of the sequences. While BLAST might identify several regions of high-scoring similarity between two sequences, FASTA only identifies the single highest scoring region.

We have already commented that the principal advantage of FASTA and BLAST over dynamic programming is that of speed. This has been very important for those developing WWW access to sequence database searches. When a database is accessed regularly by users from all over the world, it is not uncommon that the computers running the searches have several jobs running at the same time. When this happens even the most powerful computers can have problems, and the amount of computational resource required by each search takes on crucial importance. For this reason, WWW developers have used the BLAST and FASTA algorithms in preference to full dynamic programming with the Smith–Waterman algorithm.

Statistical scores   One of the main problems with sequence similarity searches is to know when an identified sequence similarity is significant, that is, when it represents a real biological or evolutionary relationship rather than a chance similarity between unrelated sequences. In Topic E2, we discussed how alignments of protein sequences could be scored using a combination of substitution matrices and gap penalties. The higher the score of an alignment by this method the more likely it is to reflect a real biological relationship, but how high is a high score?

A major breakthrough in sequence similarity searches was the development of a statistical theory of the alignment scores. Like many theories of statistical significance, this theory centers on the calculation of *p* **values**. In the sense of a sequence similarity search, the *p* value of an identified similarity of score S is the probability that a score of at least S (i.e. a score of S or greater) would have been obtained in a chance match between two unrelated sequences of similar composition and length. Very low values of *p* therefore correspond to significant matches: in these cases it is highly improbable that the score obtained occurred by chance, and it is probable that it occurred as the result of a real biological or evolutionary relationship. All the search methods we have discussed so far report *p* values for identified similarities, and these can be used to judge the significance of the discovered relationships. Of course, the user still has to form a judgment about his/her own threshold *p* value for considering a similarity significant. Clearly a value of $p=0.5$ does not represent a significant similarity (because it has a 50% probability of occurring by chance between unrelated sequences), but values of *p* less than 0.01 are much more significant.

A related quantity that is reported by most search software is the E **(Expect) value**. For an identified similarity of score S, the E value is the expected frequency of scores of at least S. It can be interpreted as the number of scores of at least s that would have been expected to have occurred by chance. The calculation of *p* and E values can be carried out for the two matched sequences in isolation (i.e. the probability of a match between the two sequences scoring at least as high by chance) or for the entire database search (i.e. the probability of a score at least as high in the entire database search). The latter are generally higher (because the database is larger than any individual sequence), and are obviously the most relevant in assessing database search results. FASTA and BLAST take slightly different approaches to whole database *p* and E values.

The E values reported by the FASTA software are related to the p value by the equation $E=Np$, where N is the number of sequences in the database (i.e. E refers to the database search but p to the sequences in isolation). This reflects an assumption that all the database sequences are equally likely to be related to the query sequence, which is questionable when they may differ significantly in length. The calculation of E values for BLAST does not make this assumption, but rather treats the database as one very long sequence. Nevertheless, the FASTA calculation illustrates the following important fact: even when the p value for a match between two sequences is low, E values for search of a large database can be quite large. For instance a FASTA similarity score S with a p value of 0.001 might seem fairly significant, but in a database of 10 000 sequences this would correspond to an E value of $0.001 \times 10\,000 = 10$. So, we would expect the scores of at least S to occur 10 times by chance, which is clearly not a high level of significance. For this reason E values are often more informative than p values.

Sensitivity and specificity

Evaluation of the results of a database search is best viewed in terms of two complementary measures, the **sensitivity** and the **specificity**. Suppose a threshold value of E or p is chosen so that sequence similarities corresponding to p or E values lower than the threshold are considered significant. It is common to call the sequences with significant similarities **hits**. The search partitions the database into two subsets, hits (positives) and non-hits (negatives). These can be further partitioned, at least conceptually, into true and false positives and true and false negatives. A true positive is a hit with a real biological relationship to the query sequence, and a false positive is a hit without such a relationship. A true negative is a non-hit with no real biological relationship to the query, and a false negative is a non-hit with a real biological relationship to the query. The **sensitivity** of the search is the proportion of the real biological relationships in the database that were detected as hits, and it can be written as

$$S_n = n_{tp}/(n_{tp} + n_{fn})$$

where n_{tp} is the number of true positives and n_{fn} is the number of false negatives. The **specificity** of the search is the proportion of hits that correspond to real biological relationships, and it can be written as

$$S_p = n_{tp}/(n_{tp} + n_{fp})$$

where n_{fp} is the number of false positives. An ideal database search would have sensitivity and specificity both as close as possible to 1, but in practice, this is not attainable. Note that there is a trade-off between the two quantities. If the threshold for significance is increased, so that more hits are considered significant, then the sensitivity is likely to increase (there will be more true positives and less false negatives), but the specificity is likely to decrease (there will be more false positives).

It is of course only possible to carry out an analysis of sensitivity and specificity if the real biological relationships in the database are already known (so that they can be assigned to the categories of true and false positive and negative). Further, this information must come from a source independent of sequence similarity analyses, for instance experimental determination of protein structure and function. In general, this information is not available, but it is useful to understand the trade-off between sensitivity and specificity when choosing threshold p and E values. Some investigators have used ideas like these with sets of proteins of known structure and function to compare the performance of FASTA, BLAST and Smith–Waterman methods, and have found evidence for slightly better performance of the Smith–Waterman by comparison with the faster methods.

Database types

The database search methods we have discussed can be used with protein or nucleic acid databases and query sequences. The names of the alternative programs are given in *Table 1*. Several of these programs first translate a nucleic acid sequence in all possible reading frames to produce six possible coded protein sequences. Sequences are then compared at the protein level rather than the nucleic acid level. When these options are chosen in preference to options that compare simply at the nucleic acid level, the searches take longer to run but the results are better. This is because it is possible to detect much more distant relationships using protein sequences, as we explained in Topic E2.

Table 1. BLAST and FASTA programs ('translated' means the nucleic acid sequence is translated to protein in six reading frames)

Program name	Query sequence	Database type
Blastp	Protein	Protein
Blastn	Nucleic acid	Nucleic acid
Blastx	Nucleic acid (translated)	Protein
Tblastn	Protein	Nucleic acid (translated)
Tblastx	Nucleic acid (translated)	Nucleic acid (translated)
Fasta	Protein or nucleic acid	Protein or nucleic acid
Tfastx	Protein	Nucleic acid (translated)
Fastx	Nucleic acid (translated)	Protein

Worked example and program availability

If you want to use BLAST or FASTA to search the large, general-purpose sequence databases like SWISS-PROT, GenBank and that at the European Molecular Biology Laboratory (EMBL) then it is probably best done using the WWW sites of the organizations that maintain these databases, that is, the European Bioinformatics Institute (EBI) and the National Center for Biotechnology Information (NCBI). Appropriate links are available on the text WWW site and in Topic A3. Here we give a worked example using FASTA at the EBI site and BLAST at the NCBI site.

The results of the worked example are shown in *Figs 1–4*. The search carried out was with the sequence of human pleckstrin and similar sequences were sought in the SWISS-PROT database. The searches were carried out with the BLOSUM62 amino acid substitution matrix (Topic E2) and gap-opening and gap-extension penalties of 8 and 2. These relatively low penalties were used to achieve maximum comparability between the two search servers using the limited sets of values available at each site, and ideally the higher default penalties of 12 and 2 should be used. *Figure 1* shows the significant similarities found by FASTA. Using 0.1 as a threshold E value below which similarity was considered significant, the algorithm found 17 significant sequence similarities. The sequence alignment associated with one of the similar sequences (human cytohesin 1, the 12th on the list of significant similarities) is shown in *Fig. 2*. Note that the alignment covers almost the entire length of the pleckstrin sequence.

Figure 3 shows the significant similarities discovered by BLAST. Note that in

Sequence			Len	opt	bits	E
SW:PLEK_HUMAN	P08567	PLECKSTRIN (PLATELET P47 PRO	(350)	1854	469	7.1e-132
SW:KRAC_DICDI	P54644	RAC-FAMILY SERINE/THREONINE	(444)	140	46	0.00029
SW:Y053_HUMAN	P42331	HYPOTHETICAL PROTEIN KIAA005	(638)	139	46	0.00045
SW:Y041_HUMAN	Q15057	HYPOTHETICAL PROTEIN KIAA004	(632)	123	42	0.007
SW:AKT3_MOUSE	Q9WUA6	RAC-GAMMA SERINE/THREONINE P	(479)	115	39	0.022
SW:AKT3_HUMAN	Q9Y243	RAC-GAMMA SERINE/THREONINE P	(479)	115	39	0.022
SW:CYH3_MOUSE	O08967	CYTOHESIN 3 (ARF NUCLEOTIDE-	(399)	110	38	0.045
SW:SPCO_HUMAN	Q01082	SPECTRIN BETA CHAIN, BRAIN (	(2364)	118	41	0.046
SW:AKT3_RAT	Q63484	RAC-GAMMA SERINE/THREONINE PRO	(454)	108	38	0.07
SW:CYH1_MOUSE	Q9QX11	CYTOHESIN 1 (CLM1).	(398)	107	37	0.075
SW:CYH1_RAT	P97694	CYTOHESIN 1 (SEC7 HOMOLOG A) (	(398)	107	37	0.075
SW:CYH1_HUMAN	Q15438	CYTOHESIN 1 (SEC7 HOMOLOG B2	(398)	107	37	0.075
SW:CYH4_HUMAN	Q9UIA0	CYTOHESIN 4.	(394)	106	37	0.088
SW:SPCO_MOUSE	Q62261	SPECTRIN BETA CHAIN, BRAIN (	(2363)	114	40	0.091
SW:RSG2_MOUSE	P58069	RAS GTPASE-ACTIVATING PROTEI	(848)	109	38	0.096
SW:RSG2_HUMAN	Q15283	RAS GTPASE-ACTIVATING PROTEI	(849)	109	38	0.096
SW:3BP2_HUMAN	P78314	SH3 DOMAIN-BINDING PROTEIN 3	(561)	107	38	0.098
SW:CYH2_MOUSE	P97695	CYTOHESIN 2 (ARF NUCLEOTIDE-	(400)	105	37	0.11

Fig. 1. Results obtained by searching the SWISS-PROT protein sequence database for sequences similar to human pleckstrin with the FASTA program using the server at the EBI. This shows the highest scoring hits. Each line shows a single protein sequence. First come database accession codes and then the name followed by the length of the sequence in parenthesis. Following that are two scores that are of technical interest only (opt and bits), and on the end of the line is the FASTA E value. Scores are given down to an E value cutoff of 0.1.

this case BLAST finds more similarities below the $E=0.1$ cutoff chosen. BLAST and FASTA use different ways of calculating E values and it is unsurprising that they produce different results. Both algorithms find the significant similarity to the human cytohesin 1 sequence featured in *Fig. 2*. The alignments from BLAST are shown in *Fig. 4*. BLAST finds two regions of similarity, but unlike FASTA does not find similarity over the entire sequence.

In the case of these sequences, it is instructive to compare the results with what is already known about the sequences. It is known that the human pleckstrin sequence contains two pleckstrin homology (PH) domains (amino acids 5–100 and 245–345) separated by another unrelated domain. Cytohesin 1 contains a PH domain (amino acids 260–375) preceded by an unrelated domain. The first of the BLAST alignments corresponds to the match between second PH domain of pleckstrin and the PH domain of cytohesin, and this similarity is also the part of the FASTA alignment where similarity is strongest. The second BLAST alignment is the similarity between the first PH domain of pleckstrin and the PH domain of cytohesin. BLAST does not discover any sequence similarity outside the PH domains. Since no similarity was expected here, it seems that this is a better result than that produced by FASTA. In this case, FASTA would have performed better with higher gap penalties, and this one case does not give sufficient evidence for any real comparison of the performance of the two methods. However, it does illustrate an important general point. The list of significant similarities ranked by E value is useful, but it is almost always necessary to inspect the sequence alignments to see where the sequences are most similar and where they are unrelated.

```
>>SW:CYH1_HUMAN Q15438 CYTOHESIN 1 (SEC7 HOMOLOG B2-1).   (398 aa)
 initn: 111 init1:  76 opt: 107  Z-score: 155.6  bits: 37.5 E(): 0.075
Smith-Waterman score: 152;   23.864% identity (31.343% ungapped) in 352 aa over
(37-347:70-378)


           10        20        30        40        50        60
PLEK_H REGYLVKKGSVFNTWKPMWVVLLEDGIEFYKKKSDNSPKGMIP--LKGSTLTSPCQDFGK
                             .:: . .:: :   .... : . :.:...
SW:CYH EIAEVANEIENLGSTEERKNMQRNKQVAMGRKKFNMDPKKGIQFLIENDLLKNTCEDIAQ
          40        50        60        70        80        90


           70        80        90       100       110
PLEK_H RMFVFK---ITTTKQQDHFFQAAFLEERDAW-VRDINKAIKCIEGGQKFARKSTRR---S
        :..:    .. :     :       .: ::: . .. ..  .. ..  .. :.   :
SW:CYH --FLYKGEGLNKTAIGD------YLGERDEFNIQVLHAFVELHEFTDLNLVQALRQFLWS
         100       110             120       130       140       150


        120       130       140       150       160       170
PLEK_H IRLP---ETIDLGALYLSMKDTEKGIKELNLEKDKKIFNHCFTGNC-VIDW--LVSNQSV
        .::: . ::        : ..  .. ..      : .: ... . :.
SW:CYH FRLPGEAQKID------RMMEA---FAQRYCQCNNGVFQS--TDTCYVLSFAIIMLNTSL
          160           170       180       190       200


              180       190       200       210       220
PLEK_H RN---RQEGLMIASSLLNEGYLQPAGDMSKSAVDGTAENPFLDNPDAFYYFPDSGFFCEE
        .:   ...  .     .:.: .. .::. . .  :. . ..:   . ..:    :.
SW:CYH HNPNVKDKPTVERFIAMNRG-INDGGDLPEELLRNLYES-IKNEP---FKIP------ED
          210       220       230       240


        230       240       250       260       270       280
PLEK_H NSSDDDVILKEEFRGVIIKQGCLLKQGHRR-KNWKVRKFILREDPAYLHYYDPAGAEDPL
        ...:   : :   .:  ::: :  : :.:: .  : ::: ..  :.:.. ..   ..:
SW:CYH DGND----LTHTFFNPD-REGWLLKLGGGRVKTWKRRWFILTDN--CLYYFEYTTDKEPR
         250       260       270       280       290       300


        290       300       310       320
PLEK_H GAIHLRGCVVTSVESNSNGRKSEEENLFEII---TADEV---------------H--YF
        : : :.. . .::. :... : ::.  . :.:       : :
SW:CYH GIIPLENLSIREVED------SKKPNCFELYIPDNKDQVIKACKTEADGRVVEGNHTVYR
         310           320       330       340       350


        330       340       350
PLEK_H LQAATPKERTEWIKAIQMA-SRTGK
        ..: ::.:. ::::.:. : ::
SW:CYH ISAPTPEEKEEWIKCIKAAISRDPFYEMLAARKKKVSSTKRH
         360                 370                 380
```

Fig. 2. The FASTA alignment corresponding to one of the high-scoring similarities (actually the 12th best score) from the search in Fig. 1. Many of the details in the header are of technical interest only, but the E value is reported along with a percentage identity for the alignment.

```
Sequences producing significant alignments:                    (bits) E Value
gi|4505879|ref|NP_002655.1|   (NM_002664) pleckstrin; p47 [Ho...   726   0.0
gi|7661882|ref|NP_055697.1|   (NM_014882) KIAA0053 gene produ...    50   5e-06
gi|7019505|ref|NP_037517.1|   (NM_013385) pleckstrin homology...    50   7e-06
gi|1730069|sp|P54644|KRAC_DICDI   RAC-FAMILY SERINE/THREONINE...     49   1e-05
gi|8392888|ref|NP_058789.1|   (NM_017093) murine thymoma vira...    47   5e-05
gi|7242195|ref|NP_035312.1|   (NM_011182) pleckstrin homology...    45   1e-04
gi|6680674|ref|NP_031460.1|   (NM_007434) thymoma viral proto...    45   1e-04
gi|4502023|ref|NP_001617.1|   (NM_001626) v-akt murine thymom...    45   1e-04
gi|6755186|ref|NP_035311.1|   (NM_011181) pleckstrin homology...    45   2e-04
gi|8670546|ref|NP_059431.1|   (NM_017457) cytohesin 2, isofor...    45   2e-04
gi|4758964|ref|NP_004753.1|   (NM_004762) cytohesin 1, isofor...    44   3e-04
gi|2498175|sp|Q15438|CYH1_HUMAN   CYTOHESIN 1 (SEC7 HOMOLOG B)...    44   3e-04
gi|7242193|ref|NP_035310.1|   (NM_011180) pleckstrin homology...    43   7e-04
gi|13124031|sp|P97694|CYH1_RAT   CYTOHESIN 1 (SEC7 HOMOLOG A)...     43   7e-04
gi|6753032|ref|NP_035915.1|   (NM_011785) thymoma viral proto...    42   0.002
gi|4885549|ref|NP_005456.1|   (NM_005465) v-akt murine thymom...    42   0.002
gi|15100164|ref|NP_150233.1|   (NM_033230) murine thymoma vir...    41   0.002
gi|13124042|sp|O43739|CYH3_HUMAN   CYTOHESIN 3 (ARF NUCLEOTID...     41   0.003
gi|13124032|sp|P97696|CYH3_RAT   CYTOHESIN 3 (SEC7 HOMOLOG C)...     41   0.003
gi|1170702|sp|Q01314|KRAC_BOVIN   RAC-ALPHA SERINE/THREONINE ...     41   0.003
gi|400112|sp|P31748|KAKT_MLVAT   AKT KINASE TRANSFORMING PROTEIN     41   0.003
gi|400144|sp|P31750|KRAC_MOUSE   RAC-ALPHA SERINE/THREONINE K...     41   0.004
gi|4885061|ref|NP_005154.1|   (NM_005163) serine/threonine pr...    40   0.005
gi|4506927|ref|NP_003014.1|   (NM_003023) SH3-domain binding ...    40   0.007
gi|6755496|ref|NP_036023.1|   (NM_011893) SH3-domain binding ...    39   0.010
gi|13928778|ref|NP_113763.1|   (NM_031575) thymoma viral prot...    39   0.011
gi|3183205|sp|Q15057|Y041_HUMAN   HYPOTHETICAL PROTEIN KIAA00...     39   0.012
gi|14777221|ref|XP_027525.1|   (XM_027525) hypothetical prote...    38   0.021
gi|6094287|sp|P91621|SIF1_DROME   STILL LIFE PROTEIN TYPE 1 (...     38   0.030
gi|465965|sp|P34512|YMX4_CAEEL   HYPOTHETICAL 43.2 KDA PROTEI...     37   0.033
gi|6094288|sp|P91620|SIF2_DROME   STILL LIFE PROTEIN TYPE 2 (...     37   0.033
gi|6755288|ref|NP_035375.1|   (NM_011245) RAS protein-specifi...    37   0.040
gi|121515|sp|P28818|GNRP_RAT   GUANINE NUCLEOTIDE RELEASING P...     37   0.041
gi|13124259|sp|Q13972|GNRP_HUMAN   GUANINE NUCLEOTIDE RELEASI...     36   0.074
gi|121742|sp|P09851|RSG1_BOVIN   RAS GTPASE-ACTIVATING             36   0.012
```

Fig. 3. Results obtained by searching the SWISS-PROT protein sequence database for sequences similar to human pleckstrin with the BLAST program using the server at the NCBI. This shows the highest scoring hits. Each line shows a single protein sequence. First come database accession codes (which differ from those in Fig. 1 for the same sequences) and then the name of the sequence. Following that is a score that is of technical interest only (bits), and on the end of the line is the BLAST E value. Scores are given down to an E value cutoff of 0.1.

```
>gi|2498175|sp|Q15438|CYH1_HUMAN CYTOHESIN 1 (SEC7 HOMOLOG B2-1)

 Score = 44.3 bits (103), Expect = 3e-04
 Identities = 36/121 (29%), Positives = 55/121 (44%), Gaps = 30/121 (24%)

Query: 247 KQGCLLK-QGHRRKNWKVRKFILREDPAYLHYYDPAGAEDPLGAIHLRGCVVTSVESNSN 305
            ++G LLK  G R K WK R FIL ++   L+Y++    ++P G I L   +  VE
Sbjct: 263 REGWLLKLGGGRVKTWKRRWFILTDN--CLYYFEYTTDKEPRGIIPLENLSIREVED--- 317

Query: 306 GRKSEEENLFE--------------------IITADEVHYFLQAATPKERTEWIKAIQM 344
             S++ N FE                     ++  +   Y + A TP+E+ EWIK I+
Sbjct: 318 ---SKKPNCFELYIPDNKDQVIKACKTEADGRVVEGNHTVYRISAPTPEEKEEWIKCIKA 374

Query: 345 A 345
            A
Sbjct: 375 A 375


 Score = 40.4 bits (93), Expect = 0.005
 Identities = 17/50 (34%), Positives = 32/50 (64%), Gaps = 1/50 (2%)

Query: 7   REGYLVK-KGSVFNTWKPMWVVLLEDGIEFYKKKSDNSPKGMIPLKGSTL 55
            REG+L+K  G    TWK  W +L ++ + +++  +D  P+G+IPL+  ++
Sbjct: 263 REGWLLKLGGGRVKTWKRRWFILTDNCLYYFEYTTDKEPRGIIPLENLSI 312
```

Fig. 4. The BLAST alignments corresponding to one of the high-scoring similarities (actually the 12th best score) from the search in Fig. 3. Query is the human pleckstrin sequence, and Sbjct is the database sequence (human cytohesin 1 in this case). This alignment is to the same database sequence as the one shown in Fig. 2. Many of the details in the header are of technical interest only, but the E value is reported along with a percentage identity (29%). Note that BLAST finds two separate regions of similarity between the sequences, the first (amino acids 247–345) scoring much more strongly than the second (amino acids 7–55).

E4 SEQUENCE FILTERS

Key Notes

Nonspecific sequence similarity

Certain types of sequence similarity are less likely to be indicative of an evolutionary relationship than others are. Examples of this are similarity between regions of low compositional complexity, short period repeats and protein sequences coding for generic structures like coiled coils.

Similarity searches

Regions of the types mentioned above can degrade the results of similarity searches and are often filtered out of query sequences prior to searching. The programs SEG and DUST can be used to detect and filter low complexity sequences, XNU can filter short period repeats and COILS can detect the presence of potential coiled coil structures.

Related topics

Sequence similarity searches (E1)
Amino acid substitution matrices (E2)

Database searches: FASTA and BLAST (E3)

Nonspecific sequence similarity

Sequences or sequence segments may be similar to each other for reasons that are not evolutionary or do not reflect similarity in biological function. Sequences with biased composition or containing short period repeats are perhaps the most obvious examples of this. In protein sequences, it is common to find short segments whose composition is dominated by a small number of amino acids, and these segments are said to have low compositional complexity. For example, part of the sequence of the human Huntington's disease protein is shown in *Fig. 1* with **low complexity regions** associated with compositional bias towards glutamine and proline. It is interesting that in this case the length of the low complexity glutamine repeat, produced by unstable trinucleotide repeat expansion, is related to the severity of the disease. It is possible that two sequences each contain low complexity segments with similar composition (e.g. another protein containing repeated glutamine and/or proline), but this type of similarity is unlikely to indicate that the proteins are homologous or that they share similar biological function. It is much more likely to reflect a common mechanism for the creation of the low complexity or repeat region, such as unstable expansion of a trinucleotide repeat.

Another type of nonspecific sequence similarity is that associated with proteins that form **coiled coil structures**. These structures comprise two alpha

```
MATLEKLMKA FESLKSFQQQ QQQQQQQQQQ QQQQQQQQQQ PPPPPPPPPP PQLPQPPPQA
QPLLPQPQPP PPPPPPPGP  AVAEEPLHRP KKELSATKKD RVNHCLTICE NIVAQSVRNS
PEFQKLLGIA MELFLLCSDD AESDVRMVAD ECLNKVIKAL MDSNLPRLQL ELYKEIKKNG
APRSLRAALW RFAELAHLVR PQKCRPYLVN LLPCLTRTSK RPEESVQETL AAAVPKIMAS
```

Fig. 1. Part of the sequence of the human Huntington's disease protein (Huntingtin) showing low complexity regions (underlined) associated with compositional bias towards glutamine (Q) and proline (P).

helices coiled around each other and are associated with a seven-residue pseudo-repeat in the amino acid sequence, in which residues one and four are predominantly nonpolar. Examples of such structures are found in keratin and myosin. It seems that this type of structure is one that is advantageous to peptide sequences in general, and has been formed by many nonhomologous proteins in a process of convergent evolution to a similar structure. The presence of a shared coiled coil structure is not strong evidence for an evolutionary relationship between two proteins.

Similarity searches

Similarity searches usually aim to detect homologous proteins or proteins with similar functions. Nonspecific sequence similarity therefore degrades the results of such searches. It is common to remove segments of the query sequence that might be associated with nonspecific sequence similarity prior to carrying out a similarity search. Several programs (which are linked from the text WWW site) are available to carry this out. The programs **SEG** and **DUST** will detect low complexity sequences. SEG is used mainly for protein sequences but will work with nucleic acids, and DUST is exclusively for nucleic acid sequences. The program **XNU** will detect short periodicity repeats. The program COILS will predict potential coiled coil structures in proteins.

Many WWW based sequence similarity search servers give the user an option to use the standard filters (SEG, XNU or DUST) prior to carrying out the search. These typically take the query sequence and replace monomers in the identified regions with the letter X so that they are ignored by the search software.

E5 ITERATIVE DATABASE SEARCHES AND PSI-BLAST

Key Notes

Detection of evolutionary relationships	Divergent evolution can change protein sequences beyond recognition while preserving structure and function. Methods like BLAST and FASTA sometimes only detect a small proportion of the evolutionary relationships in a database, and much bioinformatics research has focused on the detection of distant evolutionary relationships between sequences.
Iterative database searches	PSI-BLAST is an iterative search method that improves on the detection rate of BLAST and FASTA. Each iteration discovers intermediate sequences that are used in a sequence profile to discover more distant relatives of the query sequence in subsequent iterations. Potential problems with PSI-BLAST are associated with the potential for unrelated sequences to pollute the iterative search, and difficulties associated with the domain structure of proteins. PSI-BLAST often detects up to twice as many evolutionary relationships as BLAST.
Related topics	Sequence similarity searches (E1) Amino acid substitution matrices (E2) Database searches: FASTA and BLASTA (E3) Sequence filters (E4)	Iterative database searches and PSI-BLAST (E5) Multiple sequence alignment and family relationships (F1) Protein families and pattern databases (F2) Protein domain families (F3)

Detection of evolutionary relationships

Divergent evolution changes the sequences of homologous proteins. In long evolutionary times, these changes can be very significant, and the degree of similarity between the homologous sequences can become very small. In many cases, the degree of sequence similarity falls below the level that is detectable by methods like BLAST and FASTA. This happens when the similarity between the homologous protein sequences is similar to the level of similarity that commonly occurs by chance between unrelated sequences. In the case of proteins, we can tell that this has happened by looking at the protein three-dimensional structures, because, as we explain in Section I, structural similarity tends to be preserved even when the sequences have diverged beyond recognition. Well-known examples of this are found within the globin family of homologous proteins. These proteins share a common function (oxygen transport) and have the same structure, and yet there are pairs of members of the family from different organisms whose sequences have less than 10% identical residues.

Using the database of three-dimensional protein structures, it has been estimated that BLAST and FASTA may detect only 20% of the total number of *distant* evolutionary relationships to the query sequence within the database.

Therefore, much bioinformatics research effort has been devoted to methods that are able to improve on this detection rate. One such method is PSI-BLAST, and others are discussed in Section F.

Iterative database searches

PSI-BLAST makes use of iterated BLAST searches in order to extend the number of evolutionary relationships detected. The idea is illustrated in *Fig. 1*. An initial BLAST search is performed with the query sequence. Then, each of the hits from this search (above a chosen E value cutoff) can be used as the query sequence in a second iteration of BLAST searching. This second iteration should detect a greater number of evolutionary relationships to the initial query sequence, and this process can be repeated (iterated) until no more significant sequence similarities are found. After a number of iterations, more of the distant evolutionary relationships to the initial query sequence should have been detected. This is achieved by using information from *intermediate* sequences. The evolutionary relationship between sequences A and C may not be detectable by BLAST, but if an evolutionary relationship between A and an intermediate sequence B can be detected, along with a further relationship between B and C, then it may be inferred that A is related to C.

PSI-BLAST works in essentially this way. The only difference from the explanation above is a minor technical one. The second and subsequent iterations are not carried out as BLAST searches with each single sequence hit from the previous iteration, but rather as a BLAST search with a **sequence profile** formed from all the hits. Sequence profiles are probabilistic models of sets of related sequences that contain information from all the sequences. Using sequence profiles means that only one BLAST search is required for each iteration, rather

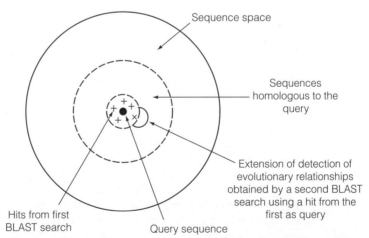

Fig. 1. The idea behind PSI-BLAST. Each sequence in the database (the sequence space) is represented by a point within the outer circle, and the query sequence is represented by the large black dot. Distances between the points representing sequences are proportional to their evolutionary distance, so that points close to each other are closely related. The sequences within the large dashed circle centered on the query sequence are all homologous to the query and are targets for detection by the similarity search. Within the smaller dashed circle, the crosses represent sequences detected as similar to the query by an ordinary BLAST search (the hits). The small circular arc attached to this circle represents further homologous sequences that could be detected by a second BLAST search, using one of the hits from the first BLAST search as query. In this case, the query sequence used is the one marked by a diagonal (St Andrew's) cross. In this way, hits from the first and subsequent BLAST searches can be used to extend the coverage of evolutionary relationships to the original query sequence.

than a BLAST search for each of the new hit sequences from the previous iteration. In practice, PSI-BLAST searches often detect up to twice as many evolutionarily related sequences as ordinary BLAST searches, which makes it a very useful tool indeed.

There is one potential problem with iterative database searches. This is contamination with unrelated sequences. If the hits from any one iteration of BLAST contain a false positive then this sequence will contaminate all further iterations with its own relatives that are not relatives of the original query. For this reason the *E* value cutoff used to choose the sequences to be employed at the next iteration of PSI-BLAST is chosen by default to be conservative, in order to minimize the risk of a false positive entering the search. A related problem that the internal workings of PSI-BLAST have to contend with is the domain structures of proteins. It is possible that an intermediate sequence (sequence B above) could be related to A by sharing a domain and to C by sharing a different

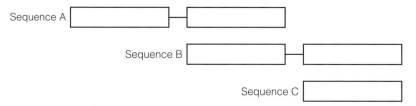

Fig. 2. *Shared domains can mean that an intermediate sequence B does not relate sequence A to sequence C. Domains are indicated by boxes. Sequence B shares its first domain with sequence A and its second with sequence C, but nothing is shared between A and C.*

domain. In such a case, there is no reason to infer that A is related to C. This is illustrated in *Fig. 2*.

A PSI-BLAST server can be found at the National Center for Biotechnology Information (NCBI) and is linked from the text WWW site. *Table 1* shows an example of the use of PSI-BLAST. In this case, the query sequence was the first pleckstrin homology (PH) domain from human pleckstrin and the objective of the search was to find other PH domains. These domains are known to be very common, but to have very divergent sequences. The parameters for the search considered a hit significant if its *E* value (Topic E3) was below 0.01, and hits with *E* values below 0.005 were used to form the sequence profile at each iteration. While the first iteration (an ordinary BLAST search) found only 93 hits, subsequent iterations added more than 500 new sequences. The search converged (no more new sequences were found) at the fourth iteration. This test case shows conclusively the advantage of PSI-BLAST over an ordinary BLAST search in detecting distant sequence similarity.

Table 1. *Detection of PH domain sequences at progressive PSI-BLAST iterations*

Iteration number	Number of PH domain sequences found
1	93
2	607
3	622
4	622

F1 MULTIPLE SEQUENCE ALIGNMENT AND FAMILY RELATIONSHIPS

Key Notes

Multiple alignment	Multiple alignment illustrates relationships between two or more sequences. When the sequences involved are diverse, the conserved residues are often key residues associated with maintenance of structural stability or biological function. Multiple alignments can reveal many clues about protein structure and function.
Software	The best-known software is the Clustal package, available by ftp from ftp://ftp-igbmc.u-strasbg.fr/pub/ClustalX.
Progressive alignment	Most commonly used software uses the method of progressive alignment. This is a fast method, but frozen-in errors mean that it does not always work perfectly. Biological knowledge can provide information about likely alignments, and where automatically produced alignments turn out to be imperfect, software for manual alignment editing is required.
Related topics	Sequence similarity searches (E1) Amino acid substitution matrices (E2) Building phyloyenetic trees (G2)	Conceptual models of protein structure (I1) The evolution of protein structure and function (I3)

Multiple alignment

Protein and nucleic sequences exist in families and their inter-relationships can be illustrated by multiple alignment of the sequences. These are like the pair-wise sequence alignments of Topic E1, but generally involve more than two sequences. Multiple alignment often tells us more than pair-wise alignment because it is more informative about evolutionary conservation. For example, two identical amino acid residues may be aligned between two protein sequences, but the fact that these have not mutated may just be down to chance. On the other hand, if a residue is conserved throughout a family of sequences that are otherwise quite diverse, then this indicates that the residue might play a key structural or functional role.

An example of a multiple alignment is shown in *Fig. 1*. This is a section of a multiple alignment of serine protease sequences, the uppermost of which is human thrombin (THRB_HUMAN). This alignment section illustrates that there are two main reasons for family-wide conservation of amino acid residues or residue properties in proteins: first to preserve function, and second to preserve structure. The biochemical function of serine proteases is the cleavage of peptide

```
SecStructure     ....................bBBBBb...----.bBBBBBb....bBBb.aaa.bba
THRB_HUMAN       LESYIDGRIVEGSDAEIGMSPWQVMLFRKSP----QELLCGASLISDRWVLTAAHCLLYP
THRB_BOVIN       FESYIEGRIVEGQDAEVGLSPWQVMLFRKSP----QELLCGASLISDRWVLTAAHCLLYP
THRB_MOUSE       LDSYIDGRIVEGWDAEKGIAPWQVMLFRKSP----QELLCGASLISDRWVLTAAHCILYP
THRB_RAT         LDSYIDGRIVEGWDAEKGIAPWQVMLFRKSP----QELLCGASLISDRWVLTAAHCILYP
LFC_TACTR        SDSPRSPFIWNGNSTEIGQWPWQAGISRWLADHNMWFLQCGGSLLNEKWIVTAAHCVTYS
FA9_RAT          EPINDFTRVVGGENAKPGQIPWQVILNGEIE------AFCGGAIINEKWIVTAAHCLK--
FA9_RABIT        QSSDDFTRIVGGENAKPGQFPWQVLLNGKVE------AFCGGSIINEKWVVTAAHCIK--
FA9_PIG          QSSDDFIRIVGGENAKPGQFPWQVLLNGKID------AFCGGSIINEKWVVTAAHCIEP-
FA7_BOVIN        NGSKPQGRIVGGHVCPKGECPWQAMLKLNGA------LLCGGTLVGPAWVVSAAHCFER-
FA7_MOUSE        NSSSRQGRIVGGNVCPKGECPWQAVLKINGL------LLCGAVLLDARWIVTAAHCFDN-
FA7_RABIT        GASNPQGRIVGGKVCPKGECPWQAALMNGST------LLCGGSLLDTHWVVSAAHCFDK-
PRTC_HUMAN       QEDQVDPRLIDGKMTRRGDSPWQVVLLDSKK-----KLACGAVLIHPSWVLTAAHCMDE-
PRTC_RAT         EELELGPRIVNGTLTKQGDSPWQAILLDSKK-----KLACGGVLIHTSWVLTAAHCLES-
PRTC_MOUSE       DELEPDPRIVNGTLTKQGDSPWQAILLDSKK-----KLACGGVLIHTSWVLTAAHCVEG-
PSS8_HUMAN       CGVAPQARITGGSSAVAGQWPWQVSITYEGV------HVCGGSLVSEQWVLSAAHCFPS-
                   : *        ***. :              *. ::    *:::****.
```

Fig. 1. *Part of a multiple alignment of some serine protease sequences. The uppermost sequence is a section of the human thrombin sequence and its secondary structure is given in the line above (a,A = helix (α, 3_{10} or π), b,B = β strand). Given in the line at the foot of the alignment are symbols indicating the degree of conservation in each column ('*' indicates a completely conserved column (same residue in each sequence), ':' indicates a column containing only very conservative substitutions and '.' indicates a column containing mostly conservative substitutions).*

bonds, as their name suggests. The cleavage is carried out by a serine (S) residue that is activated as a nucleophile by transfer of charge using neighboring histidine (H) and aspartic acid (D) residues. These three residues are known as the 'catalytic triad', and are essential to the function of the enzymes. Because they are essential to function, they are conserved in each member of the family. The conserved histidine residue is the sixth residue from the right hand side of the alignment in *Fig. 1*, and as expected, it is conserved in all the aligned sequences. Conservation of residues to preserve function is very common in proteins, but it is important to remember that there are examples of homologous proteins that have different functions. For example, as discussed in Topic I3, lysozyme and α-lactalbumin are homologous proteins with very similar sequences, but different functions (the former is an enzyme and the latter a mammalian regulatory protein).

The maintenance of a stable three-dimensional structure is another reason for residue conservation in protein families. This is discussed in detail in Section I, in particular Topics I1 and I3. There is therefore a very strong link between the conservation patterns seen in multiple sequence alignments and the underlying protein three-dimensional structure. Here we simply note some structural features in the alignment of *Fig. 1*, and refer the reader to Section I for a full discussion. The key conserved structural features in *Fig. 1* are the two conserved cysteine residues (which are disulfide-bonded to each other), the strong tendency to conserve hydrophobic residues in β-strand secondary structure elements, and the positioning of insertions and deletions outside the secondary structure elements (e.g. the four-residue insertion in the LFC_TACTR sequence).

Software

The most commonly used multiple alignment software is the **Clustal package** (see Topic O3). This software is freely available by ftp from ftp://ftp-igbmc.u-strasbg.fr/pub/ClustalX, and is linked from the text World

Wide Web (WWW) site. It works by the progressive alignment method outlined below.

Progressive alignment

The problem of multiple sequence alignment is much more difficult computationally than the pair-wise alignment described in Topic E1. Most current programs use the method of **progressive alignment**, which has the advantage of being relatively fast. This involves making a preliminary assessment of how the sequences are related using pair-wise alignments, using this to form a **guide tree** and then using this guide tree to add sequences progressively to the alignment, beginning with the most closely related sequences and finishing with the most distant. This process is illustrated in *Fig. 2*.

Progressive alignment is usually very effective, but it suffers from the problem that alignment errors made early in the process can never be rectified. They are 'frozen' into the alignment. Thus, in the example of *Fig. 1*, there might be important information in sequences C and D that could improve the alignment of A and B, but this can never be used because with the progressive algorithm A and B are aligned independently of C and D. Furthermore, independent biochemical information can sometimes give information about the correct alignment of sequences. For instance, it may be known experimentally that certain residues are key structurally or functionally and should be in conserved columns. Most often alignment errors are very obvious, such as failure to align cysteine residues involved in disulfide bonds. For these reasons it is sometimes necessary or desirable to edit multiple alignments manually. Several software products are available for this purpose, for instance the Seaview or Cinema software linked from the text WWW site.

There are several refinements often applied to the process of progressive alignment. For instance, in the Clustal suite of programs, gap penalties are varied so that gap insertion is more likely in hydrophilic loop regions, as would be expected from the discussion of the relationship of multiple alignment to protein structure above. Further, different amino acid substitution matrices (Topic E2) are applied depending on the degree of relatedness of the sequences being aligned.

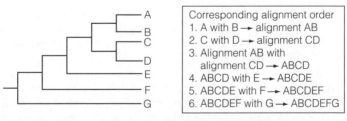

Fig. 2. Progressive alignment. The seven sequences A–G are related as shown in the guide tree on the left. This tree can be viewed as a possible way in which the sequences evolved. Lengths of the branches (horizontal lines) indicate the degree of difference between the sequences. Thus, A and B are closely related and have diverged from a near common ancestor, as are C and D. These four sequences have a common ancestor and are more closely related to each other than to sequences E, F, G, etcetera. This leads to the alignment order shown in the box on the right. First, the closely related sequences A and B are aligned, followed by C and D, and then these two alignments are aligned to form an alignment of four sequences. Sequences E, F and G are then added sequentially to this alignment. All these alignments are performed with the dynamic programming algorithms described in Topic E1.

F2 PROTEIN FAMILIES AND PATTERN DATABASES

Key Notes

Protein families	Assigning sequences to protein families is a very valuable way of predicting protein function. Many ways have been developed to represent protein family information and these have been stored in secondary protein family databases.
Consensus sequences	These condense the information from a multiple alignment into single sequence. Their main shortcoming is the inability to represent any probabilistic information apart from the most common residue at a particular position. Derivation of consensus sequences illustrates that any protein family representation is subject to bias if the set of sequences from which it was derived is biased.
PROSITE	The PROSITE database contains sequence patterns associated with protein family membership, specific protein functions and post-translational modifications. A special notation involving square brackets (e.g. [LIVM]), curly brackets (e.g. {FD}) and $x(n)$ is used to express alternative residues at each position in the pattern. The database is manually curated and any known false positives or false negatives are reported. Some of the patterns, particularly short post-translational modification patterns, suffer from a lack of specificity and occur many times in some sequences. The database also contains some sequence profile entries.
PRINTS and BLOCKS	These represent protein families by of multiply aligned ungapped segments (motifs) derived from the most highly conserved regions of the sequences. By representing more of the sequence, they have the potential to be more sensitive than short PROSITE patterns. The ability to match in only a subset of the motifs associated with a particular family means that they have the ability to detect splice variants and sequence fragments and to represent subfamilies. WWW-based search engines for the databases are available.

Related topics	Annotated sequence database (C2)	Multiple sequence alignment and
	Sequence similarity searches (E1)	family relationships (F1)
	Amino acid substitution matrices (E2)	Protein domain families (F3)

Protein families   In Topic F1, we discussed how multiple alignments of sequences from the same protein family can be used to deduce much important information about the structure, function and key amino acid residues of the family. For this reason it has been important to store information from multiple alignments of protein sequences in databases. There are many possible representations of the information within a multiple alignment, including the alignment itself, consensus sequences, conserved residues and residue patterns, sequence profiles and other probabilistic models of the sequence family. These are all useful depending on the application in mind, and most have been developed and stored in databases for large numbers of different protein families. We refer to these databases as **secondary databases**, because the information within them is not raw experimental data, which would be stored in **primary databases**, but has been derived in some way from experimental data (Topic C2).

Consensus sequences

Perhaps the simplest and most intuitive way of condensing the information in a multiple alignment is to use a consensus sequence. These can be derived in several ways, depending on the software used, but the same general idea is used in all cases. This is to produce a single sequence in which each residue is the most common, or consensus, for the sequence family. A method of derivation of a consensus sequence is illustrated in *Fig. 1*, where the consensus residue at a particular position is the most common residue in the corresponding alignment column as long as it is shared by more than 60% of the sequences. If no residue is shared by 60% of the sequences then the consensus residue is X (any residue). Consensus sequences can be useful summaries of protein families, but are less powerful than some of the methods we will discuss later, and so databases of consensus sequences are not commonly used.

Figure 1 illustrates some important general principles of protein family representation. First, it shows the essential weakness of the consensus sequence approach. This is that much information from the sequences that do not contain the consensus residue is ignored, even though these hold information about allowed substitutions at that position. It also illustrates an important consideration for all protein family representation methods. The set of sequences used in *Fig. 1* contains four thrombin sequences and two factor 9 sequences: it is therefore biased in favor of thrombin. The consensus sequence derived is also biased, being a better representation of thrombin than factor 9, and not a general representation of these two serine protease families. For instance, because the requirement is that a residue be shared by 60% of the sequences in order to be the consensus residue, it is sufficient that it appear in all four thrombin sequences, irrespective of the corresponding residues in the factor 9 sequences. This is an important point that must be considered in all family representation methods. It is very important to address the issue of possible bias in the set of sequences that have been used in the derivation, otherwise the representation derived will itself be biased.

```
THRB_HUMAN     LESYIDGRIVEGSDAEIGMSPWQVMLFRKSPQELLCGASLISDRWVLTAAHCLLYP
THRB_BOVIN     FESYIEGRIVEGQDAEVGLSPWQVMLFRKSPQELLCGASLISDRWVLTAAHCLLYP
THRB_MOUSE     LDSYIDGRIVEGWDAEKGIAPWQVMLFRKSPQELLCGASLISDRWVLTAAHCILYP
THRB_RAT       LDSYIDGRIVEGWDAEKGIAPWQVMLFRKSPQELLCGASLISDRWVLTAAHCILYP
FA9_RAT        EPINDFTRVVGGENAKPGQIPWQVILNGEIE--AFCGGAIINEKWIVTAAHCLK--
FA9_RABIT      QSSDDFTRIVGGENAKPGQFPWQVLLNGKVE--AFCGGSIINEKWVVTAAHCIK—

Consensus       XXSYIXGRIVEGXDAEXGXXPWQVMLFRKSPQELLCGASLISDRWVLTAAHCXLYP
```

Fig. 1. Deriving a consensus sequence from a multiple alignment using thrombin (THRB) and factor 9 (FA9) sequences as an example. Each position in the consensus corresponds to a column in the alignment. The consensus residue is the most common residue in the column if it is shared by more than 60% of the sequences, or X otherwise. The 60% threshold is usually variable.

PROSITE

PROSITE is a database of **sequence patterns** associated with protein family membership. It is developed by a largely manual process of seeking the patterns that best fit particular protein families and functions. For instance, two patterns are associated with the serine protease family. These are:

$$[LIVM]-[ST]-A-[STAG]-H-C$$

and

$$[DNSTAGC]-[GSTAPIMVQH]-x(2)-G-\ [DE]-S-G-[GS]-[SAPHV]-$$
$$[LIVMFYWH]\ -PA-[LIVMFYSTANQH]\ .$$

Within these patterns, square brackets indicate sets of possible residues that can occur in a particular position in sequences from the associated family. The first pattern represents a sequence of six residues and should be read as 'one of the residues L, I, V or M, followed by S or T, then A, then one of S, T, A or G, then H, then C'. In this case, the penultimate residue (H) is a key catalytic residue of the serine protease family, which was discussed in F1, and so it makes sense that there should be no alternatives at this position. The second pattern above is the sequence pattern that occurs around another key catalytic residue (serine). This pattern is longer, representing a sequence of 14 residues, and it includes an extra feature written as $x(2)$. This means two residues of any type, and not necessarily of the same type. In other patterns the x notation may be extended further, for instance $x(2,4)$ indicates that between two and four residues of any type may be present at a particular position in the family. Another extension to the notation is the use of curly brackets, for instance {LT} means that the position can be occupied by any residue *except* L or T. PROSITE patterns are most commonly derived from multiple alignments, indeed the first pattern given above is present in the multiple alignment example of Topic F1.

PROSITE patterns differ from consensus sequences in that they tend to be much shorter than the total sequence length, and that they give a means of describing a set of acceptable residues in a multiple alignment column. The patterns can be useful in assigning distant homologs to sequence families when, for instance, all that remains of the sequence similarity is limited to a few important residues around the key functional machinery of the protein. They can therefore also be indicative of shared biological function. As well as patterns associated with family membership, PROSITE contains generic patterns associated with processes like post-translational modification (glycosylation, phosphorylation, etc.). For instance the pattern

$$N-\{P\}-[ST]-\{P\}$$

is associated with *N*-glycosylation of asparagine (N). While these patterns are useful in some circumstances, they tend to be very short, and this leads to a lack of *specificity*. The pattern is likely to occur in many sequences in positions that are not real glycosylation sites. Such occurrences are known as false positives.

PROSITE patterns have a number of weaknesses. First, their shortness tends to lead to false-positive occurrences in unrelated sequences (as above), and this effect is not limited to the short patterns associated with post-translational modification. Second, while they allow the description of variation at a particular position, they have no way of attaching probabilities to the variation. For instance [LIVM] says that a position might be L, I, V or M, but it does not say that perhaps L occurs in 90% of sequences in the family and I,V or M only in the other 10%. Where PROSITE patterns have known false positives (or false negatives), they are annotated in the database. In order to overcome some of the difficulties associated with PROSITE patterns, the PROSITE database now contains **sequence profiles** in some entries. These attempt to describe longer sequence segments (usually complete domains) than the patterns, and are discussed in detail in Topic F3. There are also several other databases that are complementary tools to PROSITE in sequence analysis, including PRINTS and BLOCKS (below), Pfam and SMART (Topic F3).

The PROSITE database and various PROSITE pattern search options are available from the EXPASY World Wide Web (WWW) site (http://ca.expasy.org/prosite) and linked from the text WWW site.

PRINTS and BLOCKS

PRINTS and **BLOCKS** are closely related. Each represents protein families in terms of multiply aligned ungapped segments derived from the most highly conserved regions in a group of proteins or protein family. Such multiply aligned ungapped segments are termed **blocks** (in BLOCKS) or **motifs** (in PRINTS). In PRINTS, a set of such motifs represented a family is called a **fingerprint**. The PRINTS database is very high quality, having been created with a great deal of manual effort, and contains extensive annotation and description for the protein family and function concerned. The original version of BLOCKS was created by automatic means, but now many databases (including PRINTS) are available in BLOCKS format. An example PRINTS entry for the SH3 domain is shown in *Fig. 2*. PRINTS represents this domain by four motifs covering the most conserved areas in the multiple alignments of many SH3 domain sequences.

The motifs within these databases typically cover larger regions of the sequence than PROSITE patterns, and unlike PROSITE, matching of motifs in sequences usually takes account of amino acid substitution matrices (Topic E2), thus not requiring exact matches to a fixed pattern. For these reasons, matches to PRINTS/BLOCKS patterns are potentially more sensitive (more distant relationships can be found) and more specific (fewer false positives occur) than matches to PROSITE patterns. An advantage of the representation as a set of motifs or blocks is a natural way of representing sequence fragments and splice variants that contain only a subset of the motifs. Such sequences present more problems for the profile and hidden Markov model methods presented in Topic F3.

Search engines for the databases are available from http://bioinf.man.ac.uk/dbbrowser/PRINTS and http://www.blocks.fhcrc.org/ and are linked from the text WWW site. An example search of the PRINTS database, using

Motif 1	Motif 2	Motif 3	Motif 4
GYVSALYDYDA	DELSFDKDDIISVLGR	EYDWWEARSL	KDGFIPKNYIEMK
YTAVALYDYQA	GDLSFHAGDRIEVVSR	EGDWWLANSL	YKGLFPENFTRHL
RWARALYDFEA	EEISFRKGDTIAVLKL	DGDWWYARSL	YKGLFPENFTRRL
PSAKALYDFDA	DELSFDPDDVITDIEM	EGYWWLAHSL	YKGLFPENFTRRL
EKVVAIYDYTK	DELGFRSGEVVEVLDS	EGNWWLAHSV	VTGYFPSMYLQKS

Fig. 2. Example sequences for the four conserved motifs used to represent the SH3 domain in the PRINTS database. These motifs represent the most conserved regions from the alignment of many SH3 domains. For brevity, we show only five example sequences for each motif, but there are many more examples in the PRINTS database. Nevertheless, the conservation patterns in each motif should be clear, particularly with reference to the preferred amino acid substitutions (Topic E2).

FingerPRINTScan for a sequence fragment known to contain an SH3 domain is shown in *Fig. 3*. It can be seen that this sequence does indeed match the four motifs (*Fig. 2*) held in PRINTS for the SH3 domain with significant (low) p values. These p values can as usual be interpreted as the probability that a match scoring at least as well as the identified match would occur by chance in a random sequence.

Query sequence: YEDEEAAVVQYNDPYADGDPAWAPKNYI**EKVVAIYDYTK**DKD
DELSFMEGAIIYVIKKN**DDGWYEGVCN**R**VTGLFPGNYVESI**MHYTD

Fingerprint	Motif Number	Pval	Sequence
SH3 Domain	1 of 4	3.02e-04	EKVVAIYDYTK
	2 of 4	1.45e-06	DELSFMEGAIIYVIKK
	3 of 4	6.39e-02	DDGWYEGVCN
	4 of 4	1.25e-05	VTGLFPGNYVESI

Fig. 3. PRINTS database search. The sequence fragment above (top) was searched against the PRINTS database. A significant match to all four motifs in PRINTS representing the SH3 domain was found, and a simplified version of the output is shown. For each of the motifs the matching sequence and p value are given. The overall E value for these four motif matches was 9.0e−11 (this means 9.0×10^{-11}). Matching regions in the query sequence are shown in bold type.

F3 PROTEIN DOMAIN FAMILIES

Key Notes

Domain families	Many proteins are built up from domains in a modular architecture. The study of protein families is best pursued as a study of protein domain families. Prodom is a database of protein domain sequences created by automatic means from the protein sequence databases. The resources described in this section can be viewed as protein domain family descriptions.
Sequence profiles	These represent complete domain sequences with scores for each amino acid at each position in a multiple alignment, and position specific measures of the likelihood of insertion and deletion. They are used as an alternative to sequence patterns in some PROSITE database entries.
Hidden Markov models	These are rigorous statistical models of protein domain family sequences. They can be viewed as sequences of match, insert and delete states that generate protein sequences according to probability distributions for each state and each transition between states. A model representing a protein domain family generates sequences from that family with high probability and sequences from other families with lower probabilities. Algorithms are available to approximate the probability that a new protein sequence was generated by a particular family model, and these can be used to assign new protein sequences to families.
Resources	Pfam and SMART can be used for protein domain family analysis. The integrated resource Interpro unites PROSITE, PRINTS, Pfam, Prodom and SMART.
Related topic	Protein families and pattern databases (F2)

Domain families   In Topic F2, we discussed simple representations of conserved features in protein families, including PROSITE patterns and fingerprints. We will now move on to discuss some perhaps more sophisticated ways of describing protein families. These attempt to describe complete sequence families, including position specific insertion and deletion probabilities, rather than just the most conserved parts.

The modular architecture of proteins must be emphasized at this point. Many proteins are constructed from more than one **domain**, and some domains are common to many protein families. For instance, Src-homology (SH2 and SH3) domains appear in many proteins associated with signaling, and pleckstrin homology (PH) domains appear in many proteins that bind phospholipids. This modular architecture probably reflects the way proteins have evolved. Genetic events can result in domain swapping, domain duplication, and loss and gain of domains. Acquiring new domains with specific functions allows proteins to acquire new and more complex functions very quickly. For instance, an enzyme might gain a new domain associated with regulation of its activity, producing a protein that would be active in more-specific circumstances.

In the above, we have not defined precisely what we mean by a protein domain. This is because such a definition is difficult to construct. Domains might be defined as parts of a protein sequence with a single well-defined function (e.g. binding a particular ligand), or they might be parts of the sequence able to fold into a three-dimensional structure independent of the rest of the sequence. Equally, they might be defined just as parts of the protein three-dimensional structure that appear to be geometrically distinct. However, one important aspect of the definition of a domain is that it must be an independent unit able to exist in many, otherwise unrelated, protein sequences. Because a large number of such domains exist, it is sensible that databases of protein family descriptions should describe **domain families**. For instance, PH domains from many functionally distinct and diverse proteins may be described as a single domain family, and the structure and evolution of the domain studied independently.

The Prodom database (http://prodes.toulouse.inra.fr/prodom/doc/prodom. html), which is linked from the text World Wide Web (WWW) site, is a database of protein domain family sequences. It is created automatically from the database of known protein sequences, using pair-wise sequence comparison (as described in Section E) with the BLAST tools followed by clustering together of related sequence segments from different proteins into domain families. This procedure follows the domain definition above, in that it attempts to identify independent sequence segments with similarity detectable at the sequence level, which are shared by many proteins. These are identified with domains. The server allows a protein sequence to be compared to the database of domain sequences, in order to identify shared domains.

Sequence profiles

Sequence profiles (alternatively known as **weight matrices**) are a way of describing related sequences from a protein domain family. They have been adopted by the PROSITE database to supplement the PROSITE pattern type entries. Their principal advantage is that they describe a complete domain sequence, including the likelihood of observing each amino acid, along with the likelihood of insertion and deletion, at each position in the sequence. An example of a PROSITE sequence profile is shown in *Fig. 1*. Each multiple alignment column in the aligned family sequences is associated with a set of 20 scores, one for the appearance of each amino acid at that position. Observed amino acids tend to score highly, but the profile is derived in such a way that amino acids that substitute well for the observed amino acids score reasonably highly as well. A new sequence can be aligned to a profile so that its amino acid residues match with columns where they obtain high scores. A high total alignment score indicates that the sequence is likely to belong to the family.

Hidden Markov models

Hidden Markov models (HMMs) are the most statistically sophisticated way of representing protein domain family sequences. A schematic representation of a HMM is shown in *Fig. 2*. This comprises a set of **states**, which might be **match states** (circles labeled M), **insert states** (diamonds labeled i) or **delete states** (squares labeled d). The states are connected by arrows that represent possible **transitions** between the states. The model can be viewed as a way of generating protein sequences. Match (M) states generate amino acids according to a probability distribution for the 20 possible amino acids. For instance, a particular

```
F   K   L   L   S   H   C   L   L   V
F   K   A   F   G   Q   T   M   F   Q
Y   P   I   V   G   Q   E   L   L   G
F   P   V   V   K   E   A   I   L   K
F   K   V   L   A   A   V   I   A   D
L   E   F   I   S   E   C   I   I   Q
F   K   L   L   G   N   V   L   V   C
```

A	-18	-10	-1	-8	8	-3	3	-10	-2	-8
C	-22	-33	-18	-18	-22	-26	22	-24	-19	-7
D	-35	0	-32	-33	-7	6	-17	-34	-31	0
E	-27	15	-25	-26	-9	23	-9	-24	-23	-1
F	60	-30	12	14	-26	-29	-15	4	12	-29
G	-30	-20	-28	-32	28	-14	-23	-33	-27	-5
H	-13	-12	-25	-25	-16	14	-22	-22	-23	-10
I	3	-27	21	25	-29	-23	-8	33	19	-23
K	-26	25	-25	-27	-6	4	-15	-27	-26	0
L	14	-28	19	27	-27	-20	-9	33	26	-21
M	3	-15	10	14	-17	-10	-9	25	12	-11
N	-22	-6	-24	-27	1	8	-15	-24	-24	-4
P	-30	24	-26	-28	-14	-10	-22	-24	-26	-18
Q	-32	5	-25	-26	-9	24	-16	-17	-23	7
R	-18	9	-22	-22	-10	0	-18	-23	-22	-4
S	-22	-8	-16	-21	11	2	-1	-24	-19	-4
T	-10	-10	-6	-7	-5	-8	2	-10	-7	-11
V	0	-25	22	25	-19	-26	6	19	16	-16
W	9	-25	-18	-19	-25	-27	-34	-20	-17	-28
Y	34	-18	-1	1	-23	-12	-19	0	0	-18

Fig. 1. A PROSITE sequence profile (taken with permission from the PROSITE user manual, copyright by Amos Bairoch). A multiple alignment segment is shown above the profile. Within the profile, numbers represent scores for each of the amino acids at the corresponding position in the multiple alignment. In the first column, for instance, scores are high for the observed amino acids (F, Y, L), and also for amino acids that substitute well for these (W, V, I, M).

match state might generate large hydrophobic amino acids with high probability and others with lower probabilities. Different probability distributions apply to different match states. Insert states also generate amino acids according to probability distributions, but delete states do not generate amino acids. A protein sequence is generated by moving through the model following the arrows.

For instance, starting with the state M_0, which generates an amino acid, we could next move to any of M_1, i_0 or d_2. M_1 or i_0 would generate a second amino acid, but d_2 would not. From these states, we could move to further states in the model connected by arrows thus generating a sequence of amino acids. Note that for insert (i) states a *self-transition* is permitted, since these states have arrows linking to themselves. Just as the generation of amino acids by M and i

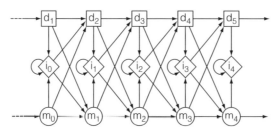

Fig. 2. A section from a HMM. Match states are labeled M, delete states d and insert states i.

states has a probability so do all the possible transitions; for instance, each of the possible transitions from M_0 (to M_1, i_0 or d_2) is associated with a probability. Thus, every path through the model, which generates a particular amino acid sequence, can be associated with a probability. This is calculated as the product of the probabilities for each component of the path, including amino acid generation probabilities from M and i states, and transition probabilities for all the arrows followed.

A HMM represents a protein domain family by generating sequences from that family with very high probability, and other sequences with much lower probabilities. The name 'hidden Markov model' is used because the sequences are generated by a Markov process, which is defined as a process in which the probability of a particular state depends only on the state immediately preceding it in a sequence. The term hidden is used because what you observe for a member of a protein family is its sequence, but hidden from you is the sequence of states (path through the model) that generated it.

You might ask what is the use of a model that generates protein sequences. Bioinformatics has enough of those already! What you really want is a way of deciding whether a particular real sequence, perhaps from a genome sequencing program, contains a domain from a particular family. The answer is that there are sophisticated algorithms that can find for a given sequence the most likely path through the model to generate it, and can use the probability of this path to estimate how likely it is that the sequence belongs to the family.

There is an obvious correspondence between the structure of the HMM in *Fig. 1* and multiple sequence alignments. The match states correspond to conserved columns in the alignment, generating amino acids similar to those observed in these columns with high probabilities. Delete states are used for sequences from the family in which the amino acid from such a column has been deleted, and insert states represent sequences with one or more inserted amino acids between the columns. The model thus describes the process of insertion and deletion of amino acids in a position-dependent way, and even gives probabilities for each possible inserted amino acid.

Resources   The **Pfam database** (http://www.sanger.ac.uk/Software/Pfam/) is a collection of protein domain family multiple alignments and HMMs. This is maintained at the Sanger Centre and can be used via a WWW interface to analyze new sequences. An example of such an analysis for the sequence of human pleckstrin is shown in *Fig. 3*. In this case, Pfam has elucidated the full domain structure of the protein, including two PH domains and a single DEP (Disheveled, Egl-10, Pleckstrin) domain. Another useful resource that will carry out this type of analysis is **SMART** (**Simple Modular Architecture Research Tool**;

Fig. 3. Pfam analysis of the human pleckstrin sequence, showing matches to the HMMs representing the N- and C-terminal PH domains and the central DEP domain. The PH domains match the pleckstrin sequences in amino acids 5–101 and 245–347, and the DEP domain matches amino acids 136–221.

http://smart.embl-heidelberg.de/). Both Pfam and SMART are linked from the text WWW site.

A recent and very valuable development is the integration of several of the protein family resources into a resource called **Interpro** (http://www.ebi.ac.uk/interpro/, also linked from the text WWW site). This unites information from many of the resources we have discussed, including PROSITE, PRINTS, Pfam, Prodom and SMART, allowing integrated search and sequence analysis.

G1 PHYLOGENETICS, CLADISTICS AND ONTOLOGY

Key Notes

Phylogenetics
Similarities and differences among species can be used to infer evolutionary relationships (phylogenies). This is because, if two species are very similar, they are likely to have shared a recent common ancestor. Many different types of character can be used in phylogenetic analysis, but nucleic acid and protein sequences are the most popular because they are common to all life forms (allowing both closely and distantly related taxa to be studied) and they can be compared objectively. However, caution must be exercised when inferring phylogenies from sequences because the rate of mutation may not be constant and sequences may be subject to differential selection.

Graphs and trees
A graph is a diagram showing relationships between particular entities, for example evolutionary relationships between species. Evolutionary relationships are generally represented by a special type of graph called a tree, which has n nodes and $n-1$ links.

Phylogenetic trees and cladograms
A phylogenetic tree is a simple way to show evolutionary relationships, with species represented by nodes and lines of descent represented by links. Phylogenetic trees may be unrooted or rooted, the latter including the position of the last common ancestor of each tree member. A phylogenetic tree that shows the evolution of species as a series of bifurcations is a binary tree or cladogram. The links in a cladogram may vary in length to convey a sense of evolutionary time.

Classification and ontologies
Classification systems are arbitrary in nature, that is, there is no standard measure of difference that defines a species, genus, family or order. Rigorous definitions are difficult to apply and such static ontologies, which are common in biological classification systems, may be replaced in the future by dynamic ontologies that are more flexible.

Related topics
Sequence similarity searches (E1)
Amino acid substitution matrices (E2)
Multiple sequence alignment and family relationships (F1)

Building phylogenetic trees (G2)
Evolution of macromolecular sequences (G3)

Phylogenetics
Living organisms are classified into groups based on observed similarities and differences. A general principle of classification systems is that the more closely related species a is to species b, the more likely they shared a recent common ancestor. In this way, similarities and differences between organisms can be used to infer **phylogenies** (evolutionary relationships). The branch of science

that deals with resolving the evolutionary relationships among organisms is **phylogenetics**.

Phylogenetics can be studied in three ways. In **phenetics**, species are grouped with others they resemble phenotypically and all characters are taken into account. In **cladistics**, species are grouped only with those that share *derived* characters, that is, characters that were not present in their distant ancestors. The third approach, **evolutionary systematics**, incorporates both phenetic and cladistic principles. Cladistics is accepted as the best method available for phylogenetic analysis because it accepts and employs current evolutionary theory, that is, that speciation occurs by bifurcation (**cladogenesis**). Information on cladistics and a useful glossary of terms can be found at the following URL: http://www.cladistics.org

Many different criteria can be used for phylogenetic analysis, including morphological characteristics, biochemical properties and, most recently, the analysis of macromolecular sequences (nucleic acid and protein sequences). Macromolecular sequences are particularly useful for comparison because they provide a large and unbiased data set, which extends across all known organisms, allowing the comparison of both closely related and distantly related taxa (Topic G3). Most importantly, however, the relatedness between sequences can be quantified objectively using sequence alignment algorithms (Topic E1). This is where bioinformatics plays an important role in phylogenetics.

The simple principle behind the phylogenetic analysis of sequences is that the greater the similarity between two sequences, the fewer mutations are required to convert one sequence into the other, and thus the more recently they shared a common ancestor. However, it is important to note that any evolutionary relationships inferred from such analysis sometimes assume a constant rate of mutation and the absence of differential selection in the sequences chosen for comparison. These conditions are seldom met! This is discussed further in Topic G2.

Graphs and trees   The clearest way to visualize the evolutionary relationships among organisms is to use a graph. In mathematics, a **graph** is a simple diagram used to show relationships between entities, such as numbers, objects or places. Entities are represented by **nodes** and relationships between them are shown as **links** or **edges** (connecting lines). Some simple graphs are shown in *Fig. 1*. These graphs can be used for a variety of purposes. For example, *G1* might represent a cyclic chemical compound with atoms at the nodes and chemical bonds as the links, and *G2* might represent a street map showing one-way streets. Note that in *G2* the links have direction (i.e. they are represented by arrows rather than simple lines). This is a **directed graph** or **digraph**.

G3 represents a special type of graph known as a **tree**. To be classed as a tree,

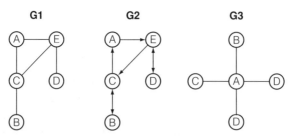

Fig. 1. Three examples of simple graphs.

a graph must have *n* nodes, *n*–1 links and no circuits. A graph contains a **circuit** if it is possible to move from one node back to itself along a sequence of links where no link is used more than once. *G1* and *G2* are not trees because each contains a circuit of nodes (A–E–C).

Phylogenetic trees and cladograms

Phylogenetic trees (also called **dendrograms**) are used to show evolutionary relationships. The nodes represent different organisms and links are used to show lines of descent. As an example, we consider the phylogenetic relationship between the entities C (chimpanzee), G (gorilla), H (human) and O (orangutan). Different trees representing this phylogeny are shown in *Fig. 2*.

The first point to note about these trees is that there are two types of node. The **ancestral nodes** (represented by boxes) give rise to branches. These may link to other ancestral nodes, or they may link to **terminal nodes** (shown as letters), which are also known as **leaves** or **tips**. Leaves represent known species and mark the end of the evolutionary pathway. Ancestral nodes may or may not correspond to a known species (e.g. the last common ancestor of humans and chimpanzees is unknown but we can infer its existence).

The second point to note is that *T1* and *T2* are **unrooted trees**, whereas *T3* is a **rooted tree**. *T1* and *T2* are identical except that *T2* is drawn in a conventional style with angled branches to look more like a real tree. These are described as unrooted trees because neither of them shows the position of the last common ancestor of all the species. In *T3*, the position of this ancestor is indicated by the node F.

The third point to note is that each tree is **binary**, that is, no ancestral nodes have more than two branches. Thus, the evolution of species is represented as a series of bifurcations, which fits in with cladistic theory. For this reason, the trees may also be termed **cladograms**.

The fourth point to note is that the length of the branches may or may not be significant. In *T1*, *T2* and *T3*, all the branches are of the same length, whereas in *T4* the branches are of different lengths. The lengths of the branches may be used to indicate the actual evolutionary distances between taxa. A cladogram which conveys a sense of evolutionary time using branch lengths may be called a **phylogram**. Note that, if *T4* represented differences between macromolecular sequences (Topic G3), there might be cases where the distance between H and C would be zero, thus 'H' and 'C' would appear on each end of a vertical line.

Finally, note that *T4* shows the same data as *T1* and *T2*, but it is presented as

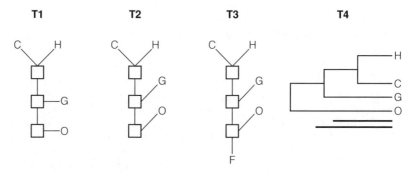

Fig. 2. Trees showing the relationship between four species, C, G, H and O (chimpanzee, gorilla, human and orangutan). Ancestral nodes in T1, T2 and T3 are represented by boxes. The thick horizontal lines under T4 are discussed in the text.

it would appear in the output format of multiple sequence alignment software, such as ClustalW/X (Topic F1). In this format, the links are still represented by lines but the ancestral nodes are represented by vertical lines rather than boxes. Phylogenetic trees such as *T4* are useful for visualizing the concept of a **clade**, which is defined as a group of organisms descended from a particular common ancestor (i.e. an ancestor and all its descendants). The groups of organisms included within a clade are defined arbitrarily. If, for example, the distance represented by the upper of the two lines beneath *T4* was thought to be significantly close, H and C would be said to be in the same clade. If, however, the criterion was the length of the lower line, H, C and G would be placed in the same clade. We explore this point in more detail below.

Figure 3 shows an example of an actual cladogram, in this case the sequence similarities among all the 12-helix integral membrane proteins known in 1996, with redundant sequences removed. In unrooted trees, there is a point referred to as the **point of trichotomy** that corresponds to the place where the ancestor would be located in a rooted tree. It is important to remember that in such a tree, the relative positions of the clades (e.g. PL, PG and PT in *Fig. 3*) are arbitrary.

Classification and ontologies

For partly historical reasons, biological science abounds with examples of hierarchical or tree-like classifications. The current version of the Linnaean system of classification into species, genera, families, orders, classes and kingdoms was the first to be developed, but the idea has been recruited into the nomenclature of the Enzyme Commission (EC numbers for enzymes) and protein structural classification systems such as SCOP and CATH (Topic I6).

The phylograms represented by *Fig. 3* and *T4* in *Fig. 2* suggest there is a method for defining a classifier objectively. That is, by using some criterion or distance, species are represented by organisms that form a clade, genera by those that form a larger clade, families and orders by those that form still larger clades, and so on. The problem that arises is that the criteria are not constant over all living organisms. For example, if the great apes in *Fig. 2* were bacteria, they would be regarded as minor variations within the same species, whereas they actually represent different families of mammals. Any attempt to overcome this problem by imposing extra rules leads to inconsistency. For example, one might say chimpanzees and humans are at least distinct species because they cannot interbreed. However, inter-species and even inter-generic crosses are commonplace among ornamental plants and among fish, so this rule cannot be rigorously applied. Among the bacteria (eubacteria), lateral gene transfer occurs widely, but does this mean that all bacteria belong to the same species?

An important point arising from the above is that classifications have to be not only objective and logical but also *useful*. It could be argued that all bacteria logically belong to the same species, but this is not useful when we want to find the causes of bubonic plague, leprosy and typhoid, or when we want to know which organisms to use in the manufacture of cheese and vinegar. The above exemplifies the problems of static ontology. **Ontology**[1] is a term used in artificial

---

[1] Do not confuse ontology with ontogeny! Phylogeny and ontogeny are often discussed together. **Ontogeny** describes the development of a particular organism, and this often reflects phylogeny (e.g. during mammalian development, embryos pass through a phylotypic stage involving the formation of gill slits, recapitulating the evolutionary pathway through a fish-like ancestor). **Ontology** refers to a systematic account of the relationship between objects, concepts or other entities, and is used in artificial intelligence and information science to structure information.

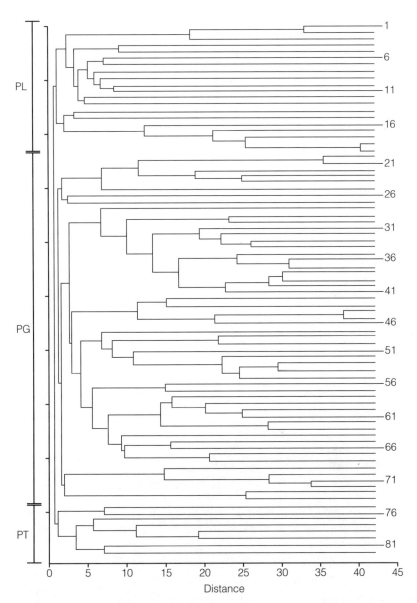

Fig. 3. Example of an actual cladogram. The leaves represent the sequences of all the 12-helix integral membrane proteins known in 1996 (with redundant members removed). The sequences are numbered arbitrarily from 1 to 82. The key is not included here. PL, PG and PT are three major clades identified by the authors. 'P' refers to the PAM matrix (Topic E3) and 'L', 'G' and 'T' refer to typical members of the clades (lactose transport, glucose transport and tetracycline resistance, respectively). This is an unrooted tree and the point of trichotomy is about half way down the PG clade.

intelligence systems to describe the relationships between entities within a given area of interest. In a **static ontology**, there is a formal and explicit specification of the relationships between entities, rather like the formal and inflexible classification systems discussed above. The future of classification may well lie in alternative ontologies, which are more flexible. A **dynamic ontology**, for example, allows the relationship between entities to be progressively refined and updated, that is, the rules can be relaxed to fit around the problem rather than forcing descriptions onto the entities themselves.

G2 BUILDING PHYLOGENETIC TREES

Key Notes

Similarity and distance tables

These are tables that show the relatedness between species for a set of chosen characters as either the percentage of matches (similarity table) or percentage of differences (distance table). Either table may be converted into a matrix and used to construct a phylogenetic tree, but distance tables are normally used for the analysis of macromolecular sequence data.

Distance matrix methods

These methods work by selecting the two most closely related taxa in a distance matrix and clustering them. The operation is then repeated until there is only one cluster left. Several variations of the method exist, which differ in the way the distances between a new cluster and other candidate taxa are calculated.

Maximum parsimony methods

In maximum parsimony methods, sequences are compared and clustered on the basis of the minimum number of mutations required to convert one sequence into another at any given position. The final phylogenetic tree is based on the overall number of changes required throughout the whole sequence.

Maximum likelihood methods

Maximum likelihood methods are similar to maximum parsimony methods but include a user-defined model that allows the probability of a given substitution occurring at any given position in the sequence to be built into the algorithm. The likelihood of a given sequence change is calculated for each position in the sequence and the most reliable tree is that with the maximum overall likelihood.

Adding a root

Some methods produce rooted trees (e.g. UPGMA) and others produce unrooted trees (e.g. neighbor joining). A root can be added to an unrooted tree either by using an outgroup, or by assuming a molecular clock.

Limitations of phylogenetic algorithms

Limitations to phylogenetic algorithms include incorrect sequence alignments, variation of evolution rates within a sequence and variation of evolution rates for the same sequence in different branches of the phylogeny. These problems can be addressed by generating robust alignments and using algorithm modifications that correct known biases. Exhaustive analysis of more than about 10 sequences is not possible because of the large number of possible trees. Heuristic and branch-and-bound methods are useful alternatives.

Phylogenetic software

Many programs are available for phylogenetic analysis. Two of the most versatile and popular are PAUP and PHYLIP. These are suitable for distance matrix, maximum parsimony and maximum likelihood analysis methods.

| How reliable are phylogenetic trees? | There is no guarantee that a given tree will accurately represent evolutionary history. However, the reliability of data can be assessed by resampling and constructing more trees, using approaches such as bootstrapping and jack-knifing. |

Related topics
Multiple sequence alignment and family relationships (F1)
Phylogenetics, cladistics and ontology (G1)

Evolution of macromolecular sequences (G3)

Similarity and distance tables

Phylogenetic trees can be constructed from either **similarity tables** or **distance tables**, which show the resemblance among organisms for a given set of characters. *Figure 1* shows an example of each type of table, where *a–e* are five species (or **taxa**) that have been scored for resemblance. As discussed in Topic G1, the characters chosen for comparison may be morphological or biochemical in nature, or may be DNA, RNA or protein sequences. If sequences are used, they are compared initially using multiple sequence alignment tools such as the ClustalW/X programs (Topic F1).

The numbers in the similarity table (*Fig. 1a*) show the *percentage of matches*, thus the diagonal (in which each species *a–e* is compared to itself) consists of 100% values. Such data form the basis of **Adansonian analysis** or **numerical taxonomy**. The numbers in the distance table (*Fig. 1b*) show *percentage differences or distances*, thus the diagonal consists of 0% values. Although either table is suitable for phylogenetic tree building using essentially the same methods, macromolecular comparisons are usually recorded as differences, so we shall use distance tables in the discussion below. A common measure of difference between macromolecular sequences is $100-S$, where S is the percentage of identical monomers when the sequences have been optimally aligned (Topic E1).

(a)

	a	b	c	d	e
a	100	65	50	50	50
b	65	100	50	50	50
c	50	50	100	97	65
d	50	50	97	100	65
e	50	50	65	65	100

(b)

	a	b	c	d	e
a	0	6	11	11	11
b	6	0	11	11	11
c	11	11	0	2	6
d	11	11	2	0	6
e	11	11	6	6	0

Fig. 1. Hypothetical (a) similarity table and (b) distance table for five organisms, a–e.

Distance matrix methods

Some of the most commonly used methods for tree building in phylogenetic analysis involve **agglomerative hierarchical clustering** based on **distance matrices**. The essential basis of this type of algorithm is that the taxa represented in a distance table (such as that shown in *Fig. 1b*) are transformed into a series of **nested partitions** by merging two taxa together in each step until only one cluster remains.

This process can be formally expressed as follows:

● Begin by creating *n* **singleton clusters**, that is, clusters containing one taxon from the distance table. For this purpose, a distance table is regarded as an

$n \times n$ matrix, with n representing the number of taxa being compared. In *Fig. 1b*, $n=5$, and the five singleton clusters are a, b, c, d and e.

- Then, determine the differences between each possible pair of clusters. This is formally expressed as a **distance function**, dis(i,j), where i and j represent any given pair of clusters. These differences are the data in the difference table shown in *Fig. 1b*.
- Next, select the pair of clusters for which dis(i,j) is minimal, that is, select the two most similar clusters. In *Fig. 1b*, this would be c and d.
- Then merge these data into a new cluster (ij). In *Fig. 1b*, this would mean defining a new cluster (cd) as the union of c and d. In the resulting phylogenetic tree, a new ancestral node would be defined with c and d as branches.
- The number of clusters in the matrix is now reduced by one. Repeat the analysis until there is only one cluster left. This is a **recursive process**, which must be carried out n–1 times.

There are a number of alternative distance matrix algorithms, which differ in the way that the distance between a new (merged) cluster and the remaining taxa in the matrix are calculated for the purpose of recursive searching. The four popular variations of the method are single linkage, complete linkage, average linkage and the centroid method. In the **single linkage** method, the distance between the merged cluster (ij) and any candidate taxon k is minimized, that is, the smaller of dis(ik) and dis(jk) is chosen. In the **complete linkage** method, the distance is maximized, that is, the larger of dis(ik) and dis(jk) is chosen. There are two variations of the **average linkage** method, in which the distance between the merged cluster (ij) and candidate taxon k is taken as the arithmetic mean of dis(ik) and dis(jk). In the **unweighted pair group method using arithmetic mean (UPGMA)**, the distance is calculated as a simple average because each candidate is weighted equally. In the **weighted pair group method using arithmetic mean (WPGMA)**, the clusters are weighted according to their size, that is, so that the candidate taxon k is equivalent in weighting to all previous taxa in the cluster. The calculations for these algorithms are therefore slightly more complex than are those for the single and complete linkage methods. Similarly, there are also weighted and unweighted variations of the **centroid method**, in which the centroid value is used rather than the arithmetic mean. These variations are known as the **unweighted** and **weighted pair group methods using centroid value (UPGMC, WPGMC)**.

The UPGMA method discussed above is popular because of its simplicity, but it makes trees with two important assumed properties that are particularly important when trees are made from macromolecular sequence data. First, it is assumed that evolution occurs at the same rate on all tree branches (this is known as the assumption of a **molecular clock**), and second it is assumed that distances in the trees are additive. The additive assumption is that the distance between any two leaves is the sum of distances on edges connecting them. As a consequence of these assumptions, the method can create incorrect trees. For example, two sequences might be very similar not because they have a direct common ancestor, but simply because they are evolving very slowly by comparison with other sequences being analyzed. UPGMA would produce a tree in which they had a direct common ancestor.

Neighbor joining (NJ) is a clustering method related to UPGMA that is able to solve some of the problems discussed above. In particular, it does not make the assumption of additivity. It is also quite fast computationally, and so is almost always a better choice of method, and is used in the ClustalW/X programs (Topic F1) to estimate trees from multiple sequence alignments. To start building an NJ tree all taxa are placed in a star, that is, individually joined to a single central node or hub through n spokes. From this star, the two taxa with the greatest similarity are chosen and are connected to a new internal node; these taxa are then known as **neighbors**. The process is repeated until the whole star is resolved into a tree. For an unrooted tree there will be $n-3$ internal branches and the process must be repeated $n-3$ times. ClustalW/X also offers bootstrapping to estimate the robustness of the generated trees (see below).

Maximum parsimony methods

In distance matrix methods, all possible sequence alignments are carried out to determine the most closely related sequences, and phylogenetic trees are constructed on the basis of these distance measurements. As an alternative, **maximum parsimony methods** can be used, in which trees are constructed on the basis of the *minimum number of mutations* required to convert one sequence into another. In proteins, this is achieved by multiple sequence alignment followed by the identification and analysis of corresponding positions in each sequence. For each aligned residue, the minimum number of base substitutions required to convert one amino acid into another is calculated. The final tree is generated by grouping those sequences that can be interconverted with the smallest number of overall changes. This method is very attractive intellectually, but, like the maximum likelihood method below, can be expensive in computer time, so NJ (above) is often to be preferred.

Maximum likelihood methods

Maximum likelihood methods also involve multiple sequence alignment and the analysis of changes at each position of the sequence. However, the difference between maximum likelihood and maximum parsimony is that the former incorporates an *expected model* of sequence changes, which weights the probability of any residue being converted into any other. This model can be set by the experimenter. For each possible tree, the likelihood of different sequence changes at each position is calculated, and these values are multiplied to provide an overall likelihood for each tree. The most reliable tree is that with the maximum likelihood.

Adding a root

Some methods automatically generate rooted trees. For instance, the clustering methods described above place the root between the final two clusters to be joined before the algorithm terminates. On the other hand, NJ produces unrooted trees; there are two main ways in which roots can be added to unrooted trees. The easiest is to use an *outgroup*. For instance, if the tree had been generated from a mixture of mammalian and bacterial sequences then it is clear that the root should lie between these two groups so that the first divergence in the tree is to divide bacteria and mammals. In this case one of the distantly related groups (e.g. the bacterial sequences) is known as the *outgroup*. In the case where there is no obvious outgroup in the sequences, a root can be added half way between two most distantly related sequences, essentially assuming a molecular clock.

Limitations of phylogenetic algorithms

All clustering methods suffer from three major limitations: incorrect sequence alignments, failure to account for variation of evolution rates at different sites within a sequence, and failure to account for sequences evolving at different rates in different taxa. These limitations often generate incorrect trees through a process known as **long branch attraction**, in which rapidly evolving sequences are grouped even if their relationship is very distant.

The problems can be addressed in several ways. The first is to make sure that sequence alignments are robust. If many diverse sequences are included in the analysis and the alignment contains many gaps, this could be a source of error. It is much better to eliminate outliers before commencing the analysis because each clustering step is definitive and cannot be undone later. Caution should be exercised in particular if a newly built tree disagrees with others generated through the analysis of different genes or proteins.

Improvements and modifications to the clustering algorithms have also made tree building much more accurate. For example, the **Farris transformed distance method** is a useful preliminary to UPGMA and the **Fitch–Margolish method** is a robust modification of WPGMA. The **paralinear distances algorithm (LogDet)** addresses the problem of unequal evolution rates and is now incorporated into many phylogenetic software packages (see below).

It is important to understand the limitations of computational power in phylogenetic analysis. Essentially, the objective of any tree-building experiment is to select the correct tree from many incorrect trees. Assuming all other limitations have been overcome, how many trees is it necessary to build to get the correct one? This depends on the number of data points being analyzed. For example, if there are three sequences, there are three possible rooted trees and one possible unrooted tree. If there are five sequences, there are 105 possible rooted trees and 15 possible unrooted trees. If there are seven sequences, there are 10 395 possible rooted trees and 954 possible unrooted trees. **Exhaustive tree search** methods, where all possible trees are created and tested, can therefore only be used for up to about 10 sequences. If more sequences need to be compared, alternative methods must be used. For example, **branch-and-bound analysis** ignores families of trees that cannot possibly give a better answer than a tree that already been found. **Heuristic analysis** samples trees randomly and can be used for many sequences, but the best tree may be missed.

Phylogenetic software

A large number of software packages are available, some free over the Internet, for the phylogenetic analysis of macromolecular sequences. Some popular programs, such as **PAUP (phylogenetic analysis using parsimony)** and **PHYLIP (phylogenetic inference package)** are versatile and allow distance matrix, parsimony and maximum likelihood method analysis to be carried out. Such packages are frequently updated with the most recent modifications and corrections to the phylogenetic algorithms. Other packages, such as **MacClade** are useful for tree manipulation. A comprehensive resource for phylogenetic software can be found at the following URL: http://evolution.genetics. washington.edu/phylip/software.html

How reliable are phylogenetic trees?

There is no guaranteed way to verify that a phylogenetic tree represents the true path of evolutionary change. However, there are ways in which to test the reliability of phylogenetic predictions. First, if different methods of tree construction give the same result, this is good evidence that the tree is reliable. Second, the data can be resampled to test their statistical significance. In a technique called **bootstrapping**, data are randomly sampled from any position within a multiple sequence alignment, and are built into new artificial alignments, which are then tested by tree building. Since the sampling is random, some positions may be sampled more than once and others not at all. Ideally, the trees built by bootstrapping should always match the original tree, and this would be defined as '100% bootstrap support'. In reality, bootstrap support of 70% or more for any given branch of a tree is taken to provide 95% confidence that the branch is correct. **Jack-knifing** is a similar process in which about 50% of the original data are resampled and used to make a new matrix, from which phylogenetic relationships are reconstructed.

G3 EVOLUTION OF MACROMOLECULAR SEQUENCES

Key Notes

Molecular phylogeny	DNA accumulates mutations over evolutionary time, leading to divergence in DNA, RNA and protein sequences in different lines of descent. This principle can be used to construct phylogenetic trees. Due to differences in intrinsic mutation rate and selective constraints, different macromolecular sequences evolve at different rates, allowing the phylogenetic analysis of both closely related and distantly related organisms.
Choice of macromolecular sequences	The macromolecular sequence chosen for a particular evolutionary study reflects the biological diversity under investigation. For closely related organisms, a rapidly evolving molecule such as mitochondrial DNA is suitable. For more diverse phylogenies, a highly conserved molecule such as ribosomal RNA is needed. Caution is required in interpreting apparent phylogenies from poorly selected macromolecules.
Rapidly evolving macromolecular sequences	Some of the most rapidly evolving sequences are found in RNA viruses, whose replication enzymes are highly error-prone. Human immunodeficiency virus is an example, where up to 30 genotypes can evolve during a single infection. This is a strategy to avoid the host's immune system.
Related topics	Multiple sequence alignment and Building phylogenetic trees (G2) family relationships (F1) Phylogenetics, cladistics and ontology (G1)

Molecular phylogeny

The DNA of organisms in different lines of descent accumulates mutations over evolutionary time leading to divergence in **macromolecular sequences** (DNA, RNA and protein sequences). Phylogenetic trees based on differences among macromolecular sequences are known as **molecular phylogenies**. Generally, the greater the divergence between two sequences the more ancient their **last common ancestor (LCA)**, so evolutionary trees can be reconstructed on this principle. Molecular phylogenies are very informative compared to those based on traditional anatomical or morphological characters. This is because they are wider in scope (e.g. it is possible to compare flowering plants and mammals using protein sequences, but not using morphological characters). There are also many different sequences to choose from (i.e. a wide range of characters), and data handling is consistent and objective.

Different macromolecular sequences evolve at different rates, even sequences in different regions of the same molecule. This is generally due to variation in **selective constraints**. Residues in an RNA or protein that have a critical structural or functional role in the molecule can accommodate mutations less easily than those in other regions. The rate at which a particular sequence evolves therefore depends largely on the proportion of residues whose substitution would adversely affect normal structure and function. As discussed below, this allows both closely related and distantly related organisms to be studied using the same methods.

Choice of macromolecular sequences

The choice of macromolecular sequence for evolutionary studies is dependent on the biological diversity under investigation. The problem is illustrated in *Fig. 1*, which shows two examples of phylogenetic trees each representing three groups of organisms. The trees are qualitative in that the distances shown are not proportionally correct, but the relative relationships between these groups are in accord with current knowledge. The main difference between the trees is the implied time span from the root to the leaves. In *Fig. 1a*, which shows three groups of primates, this is less than 10^7 years. In *Fig. 1b*, which shows the three major forms of life on this planet, the time span is approximately 4×10^9 years.

In each case, different problems have to be addressed. For the primates shown in *Fig. 1a*, the challenge is to find molecules that evolve quickly enough and therefore differ sufficiently among these closely related species. Conversely, if we wish to study the evolutionary relationship between the three groups of life shown in *Fig. 1b*, highly conserved macromolecules that tolerate few mutations are required.

A useful macromolecular sequence for the study of primates is **mitochondrial DNA (mtDNA)**. As a consequence of respiratory metabolism, there is a higher concentration of active oxygen species (such as superoxide and the hydroxyl radical) in the mitochondria than in the nucleus and consequently a higher chance of oxidative chemical lesions in mitochondrial DNA. Furthermore, the mitochondrial DNA polymerase is more error-prone than the nuclear enzyme. Therefore, mtDNA evolves more quickly than nuclear DNA due to an increased **intrinsic mutation rate**. There is a short noncoding region in primate mtDNA where selective constraints are low, since point mutations tend not to affect mitochondrial function. This particular sequence evolves at a suitable rate to study primate phylogeny. The tree shown in *Fig. 1a* is consistent with the align-

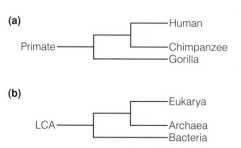

Fig. 1. *Rooted trees for (a) three great apes with an unspecified primate ancestor, and (b) the three major forms of life on this planet. Archaea were previously called the archaebacteria, Bacteria were previously called the eubacteria and by Eukarya we refer to the nuclear–cytoplasmic system in eukaryotes (organelles are ignored). LCA is the last common ancestor of all life on this planet.*

ment and clustering of this region and, to a lesser but significant extent, with such analyses of coding genes in mtDNA.

For the study of more divergent taxa, we need a ubiquitous molecule that is highly conserved in all living organisms (including animals, plants, fungi, bacteria, extremophiles and parasites). Such a molecule would be functionally constrained at almost all residues. It would therefore have a low tolerance for mutations and would evolve very slowly. **Ribosomal RNA (rRNA)** is such a molecule. The abundant secondary structure of rRNA insures that the rate of evolutionary change is slow, since compensating base changes are required in double helical regions. The tree shown in *Fig. 1b* is consistent with the alignment and clustering of this molecule, and the conclusions are compatible with those of other macromolecular studies, as summarized below:

- The division between the Archaea (archaebacteria) and the Eukarya (eukaryotes) is less deep than that between their common node and the Bacteria (eubacteria). Note also that rRNA trees have proven useful not only for such evolutionary studies; they also provide an objective and useful tool in the difficult problem of eubacterial taxonomy.
- Mitochondria and chloroplasts are derived from prokaryotes. The relationship between the chloroplasts from certain plants and individual cyanobacteria (blue green bacteria) is particularly close.
- Ribosomal RNA sequences are not going to help us to define the LCA of all life (*Fig. 1b*). The search for this organism (which is unlikely to occur in the fossil record) will rely on protein sequences and reconstructions of ancient metabolism.

Having commented on good examples of molecules suited to phylogenetic analysis, it is important that the reader understands that there are some molecules that are not well suited. One example is the confusion of orthologs (homologous genes with identical functions in different organisms) and paralogs (homologous genes with different but perhaps related functions). *Figure 2* shows two trees each generated from hemoglobin sequences taken from human, chimpanzee and horse. The left-hand tree contains both α and β globin chains from all species and is easily understood. For each chain, the phylogeny follows the accepted organism phylogeny, with chimpanzee more closely related to human than horse. The fact that the α globins are more closely related to each other than they are to the β globins is evidence that these genes were created by a duplication that happened before the divergence of human, chimpanzee and horse as separate species. The different sequences in the tree have thus been created by two processes, speciation and duplication, and in this case interpretation is easy. The right hand tree is created from the chimpanzee α chain and the β chain of the two other organisms. This does not agree with the accepted organism phylogeny. In fact, if we did not know about the α and β globin chains and had mistakenly thought that the three sequences were orthologs, we might have been tempted to deduce that the horse was our closer relative! The hemoglobin chains are well understood, but other gene families are often not. Very confusing trees can result from sequences that are mixtures of orthologs and paralogs.

Rapidly evolving macromolecular sequences

The phylogenetic trees shown in *Fig. 1* extend over millions or billions of years. However, some genetic systems evolve at a much faster rate. RNA viruses are genetically unstable because their replication mechanism is highly error-prone.

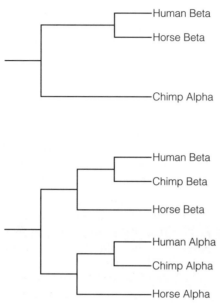

Fig. 2. Two trees generated from hemoglobin sequences from human, chimpanzee and horse. The lower tree is correct, indicating the correct phylogeny for both α and β hemoglobin chains. The upper tree is confusing, because it is formed from human and horse β chains and the chimpanzee α chain, creating the impression that horse is closer to human than chimpanzee.

Certain RNA viruses have adopted rapid evolution as a means of evading the host's response to infection. For example, **human immunodeficiency virus (HIV)** is a retrovirus of the lentivirus subgroup. Mutations arise in HIV during the course of its infection of a single individual. HIV sequences isolated from a single patient following a single infection represent a mixture of up to 30 genotypes and the course of the mutations can be followed from phylogenetic trees, in this case implying a time scale of as little as 10 years.

H1 PRINCIPLES OF GENOME ANNOTATION

Key Notes

Annotation	In bioinformatics annotation means obtaining useful information, that is, the structure and function of genes and other genetic elements, from raw sequence data. Due to differences in gene structure and genome organization, the annotation of prokaryote and eukaryote genomes involves different problems.
Finding genes by computer	Computers can be used to predict where the genes are in genomic DNA. This is achieved by a combination of signal sensing (looking for conserved motifs), content sensing (looking for regions with gene-like sequence context) and homology searching (looking for regions that match the sequences of previously discovered genes). However, no gene-finding method is 100% reliable.
Detecting open reading frames (ORFs)	In bacterial genomes, genes are rarely interrupted by introns. Therefore, a useful way to detect genes is to carry out a six-frame translation of the genome sequence and identify long ORFs. This method detects most genes but more sophisticated approaches are required to detect short genes, genes that use variations of the genetic code, and overlapping genes.
Detecting exons and introns	In higher eukaryote genomes, genes are widely dispersed and interrupted by large and numerous introns. Exons are too short to identify by ORF searching alone, so a combined approach involving exon detection by content sensing, the identification of splice signals, and the incorporation of auxiliary information such as cDNA sequences, is required to build a whole-gene model. Even so, many genes remain undetected or incorrectly delimited, and nonfunctional full or partial gene copies (pseudogenes) may be incorrectly identified as genuine genes.
RNA genes	Special methods are required to identify RNA genes since these do not contain a coding region. Such methods are based on homology searches and the prediction of secondary structure.
Annotating the human genome	The number of genes in the human genome has been the subject of much study and debate. Since the publication of the genome sequence it seems likely the number is substantially smaller than some of the estimates, which were 120 000 or more. A more likely figure is 45 000–50 000.

Related topics	Genome and organism specific databases (C3)	Sequence similarity searches (E1)
	Miscellaneous databases (C4)	Annotation tools and resources (H2)

Annotation

The term **annotation** means obtaining useful biological information from raw sequence data. Essentially this means finding genes and other functional elements in genomic DNA (**structural annotation**) and then assigning functions to these sequences (**functional annotation**). Since it became possible to sequence whole genomes, deriving biological meaning from the long sequences of nucleotides has been a crucial biological research problem. The methods discussed in Sections E and F are applied to all genome sequences for functional annotation. In this and the following topic, we consider the impact of bioinformatics on structural annotation.

A point that must be emphasized early on is that annotation involves different problems in prokaryote and eukaryote genomes. In prokaryotes, the gene density is high (i.e. there is little intergenic DNA) and the vast majority of genes have no introns. In eukaryotes, gene density falls and gene complexity increases as the organism itself becomes more complex. In the yeast *Saccharomyces cerevisiae*, for example, about 70% of the genome is made up of genes, and most of these have no introns; about 5% of *S. cerevisiae* genes have a single, small intron. In *Drosophila*, only 25% of the genome is made up of genes. About 20% of genes have no introns, and most genes have between one and four introns. In mammals and higher plants, gene density drops to 1–3%. Only 6% of human genes have no introns, and there are approximately equivalent numbers of genes with between one and 12 introns. A few human genes have more than 50 introns! In the human genome, exons are typically about 150 bp long but introns are more likely to be in the range 1–5 kb. For this reason, finding genes in higher eukaryotic genomes can be very difficult.

Finding genes by computer

There are many hybridization-based techniques and other experimental methods that can be used to detect genes in genomic DNA. However, these can only be applied on a clone-by-clone basis and cannot possibly keep pace with the rapid accumulation of genomic sequence data. Bioinformatics provides a way to fill this information gap by rapidly mining sequence data to identify potential genes. Note the use of the qualifier 'potential'. No computer-based gene-finding method is 100% accurate, and each annotation should be tested in the laboratory.

Many computer algorithms have been developed to annotate genomic sequence data and some of these are discussed in the following topic. But what features of genes do these programs identify that lets them discriminate between genes and the surrounding DNA? Essentially, three types of feature are recognized: signals, contents and homologies.

Signals are discrete, local sequence motifs such as promoters, splice donor and acceptor sites, start and stop codons, and polyadenylation sites. Such motifs tend to have **consensus sequences** and algorithms can be developed to search for these either singly or in the context of other surrounding signals. Such algorithms are known as **signal sensors**. The most sensitive of these employ weight matrices that assign a cost to each of the four nucleotides when they occur at any given position in the motif. The overall score generated by a weight matrix sensor reflects the sum of all costs at all positions, and there will be a threshold value that determines whether the candidate motif is a real signal.

Contents are extended sequences of variable length. The most important content of a gene is usually the coding region, although other sequences may be relevant, such as **CpG islands** in vertebrate genes. Contents do not have consensus sequences, but they do have conserved features that distinguish them from surrounding DNA. For example, **nucleotide frequencies** and **nucleotide dependencies** (the likelihood of two particular nucleotides occurring at two particular positions at the same time) differ between exons and introns or intergenic DNA. This can be tested by **content sensor** algorithms using statistical models.

Homologies are matches to known genes. With the large amount of sequence data available, especially expressed sequence tags (ESTs), which represent genes (Topic B1), many programs now incorporate database homology searches (Topic E3) as part of the gene-finding algorithm.

The most effective programs incorporate elements of signal and content sensing as well as homology searching to produce an integrated gene-finding package. We discuss these programs in more detail in Topic H2.

Detecting open reading frames (ORFs)

Protein-encoding genes have an **open reading frame (ORF)**, a long series of **sense codons** bracketed by a **start** or **initiation codon** (usually but not exclusively ATG) and a **stop codon** (or **termination codon**, or **nonsense codon**), which may be TGA, TAG or TAA. In bacterial genomes, ORFs are generally easy to detect because they are uninterrupted by introns. The normal procedure is to carry out a so-called **six-frame translation** (i.e. a translation of the genomic sequence in all six possible reading frames, three forwards and three backwards) and identify the longest ORF in the six possible protein sequences. Long ORFs tend not to occur by chance, so the existence of 300 bp or more of uninterrupted coding sequence in any one reading frame is good evidence for a gene.

While this method is generally sound, some genuine genes may be overlooked completely or the boundaries incorrectly specified. This applies to genes that are shorter than 300 bp (or any other length criterion applied) and genes that use rare variations of the genetic code. The standard genetic code[1] is shown in *Table 1*. In most bacterial genes, the initiation codon is ATG, but in some instances the codons GUG or UUG may be used. The frequency of alternative initiation codon usage varies in different bacterial species, so the results for any new bacterial genome sequence cannot be anticipated. ATG is also used as an internal codon, so the misidentification of an internal ATG as a start codon, especially where GUG or UUG is the genuine start codon, may lead to gene truncation from the 5′ end. In such situations, the genuine start site may be indicated by the presence of other signals, such as the **Kozak sequence** (ribosome binding site). Another example involves the stop codon TGA. In the vast majority of genes, TGA means stop, but in a very few it leads to the incorporation of the rare amino acid **selenocysteine**. This is dictated by the context of the sequence surrounding the codon, and if not recognized by a gene-finding algorithm, could lead to gene truncation from the 3′ end.

Other quirks of nature that may be missed by gene-finding algorithms are **intentional frameshifts**, which allow the production of two or more proteins from the same gene by reading through a termination codon, and the existence of **shadow genes** (overlapping ORFs on opposite DNA strands). Content sensors, which detect differences in nucleotide frequencies and dependencies, are essential for finding these genes. It is important to note that the genetic code is **degenerate** (i.e. most amino acids are specified by more than one codon) and

---

[1] Note that some bacteria and fungi use modifications of the standard genetic code, and the code also varies in mitochondria. These modifications must be taken into account when annotating, for example, mitochondrial genomes.

Table 1. The standard genetic code, tabulated to emphasize degeneracies and ambiguities

Group	Amino acids	Codons	Number	Ambiguities
1	L	CUN, UUR	6	UUG initiation (M)
	R	CGN, AGR	6	None
	A	AAN	4	None
	G	GGN	4	None
	P	CCN	4	None
	S	AGY	4	None
	T	ACN	4	None
	V	GUN	4	GUG initiation (M)
2	I	AUY, AUA	3	None
	*	UAY, UGA	3	UGA internal (Z)
	C	UGY	2	None
	D	GAY	2	None
	E	GAR	2	None
	F	UUY	2	None
	H	CAY	2	None
	K	AAR	2	None
	N	AAY	2	None
	Q	CAR	2	None
	Y	UAY	2	None
3	M	AUG	1	Initiation
	W	UGG	1	None
	Z	UGA	1	*

N is any nucleotide, Y is any pyrimidine and R any purine. Z is the amino acid selenocysteine and * represents a stop codon. Group 1 amino acids show complete degeneracy at the third position of the codon. Group 2 amino acids show partial degeneracy (purine or pyrimidine degeneracy) at the third position of the codon. Group 3 amino acids have specific codons.

that different species show different **codon bias** (i.e. a preference for particular codons). Content-sensors must therefore be trained using the **codon frequency tables** appropriate for each species.

Detecting exons and introns

In higher eukaryotes, the detection of protein coding sequences is complicated by the presence of introns, which may interrupt the ORF one or more times. Introns are best described as unhelpful to genome annotation. They may separate exons neatly, that is, leaving individual codons intact, but they may also split a codon between the first and second or second and third positions. Exons are generally small (average size 150 bp) which means that they cannot be identified simply by the detection of a long uninterrupted series of sense codons. The annotation of eukaryotic genomes therefore involves a more complex process of predicting exons using content sensors, predicting the positions of splice donor, acceptor and branch sites to identify introns and building these into a **whole-gene model**. The use of auxiliary information, for example cDNA sequences and the presence of promoter elements, is much more important for the correct identification and delimitation of eukaryotic genes than it is for prokaryote genes. The situation in eukaryotes is further complicated by the fact that many genes undergo **alternative splicing** (so that gene predictions based on cDNA sequences can miss out exons) and that eukaryotic genomes are littered with pseudogenes and gene fragments, which resemble genuine genes but are not expressed. Further data, for example from expression or mutational analysis, is often required before a gene can be confirmed as genuine.

RNA genes

The methods discussed above are suitable for the detection of genes that are translated into protein. However, a number of genes function only at the RNA level. These include **ribosomal RNA (rRNA)** and **transfer RNA (tRNA)** genes, and specialized genes such as *XIST*, which is involved in mammalian X-chromosome inactivation, and *H19*, which is involved in genomic imprinting. How are these genes identified?

Generally, the detection of **RNA genes** relies on homology searching. Such genes tend to be highly conserved because the RNAs have a defined tertiary structure, which requires extensive intramolecular base pairing. There are also specialized algorithms for detecting RNA genes, such as tRNAScan-SE, which search for sequences with the potential to form secondary structures such as stem-loops and hairpins.

Annotating the human genome

The number of protein coding genes in the human genome was always a source of intense speculation and discussion right up to the publication of the first draft complete genome sequence early in 2001. Clearly the question of how many genes are needed to make an organism as complex as a human being is very interesting. Many estimates were made, some *ad hoc*, others based on statistical consideration of EST sequences and well annotated parts of the genome that became available prior to the publication of the sequence. These estimates ranged from 30 000 to 120 000 or more, with the more popular estimates between these two limits.

On the publication of the genome sequence, it became clear that although some computational methods (see the next topic) predicted more than 120 000 genes, the number of genes present and confirmed by EST or existing protein sequence data was actually quite small, only around 30 000. It is possible then that the human being has relatively few genes. This number is to be compared with 4000 protein-coding genes in the bacterium *Escherichia coli* and around 20 000 in the nematode *Caenorhabditis elegans*. It seems likely that biological complexity is achieved by means other than gene number, for instance complex gene interactions, alternative splicing and post-translational modification.

At the time of writing, the figures on the Ensembl World Wide Web (WWW) site (http://www.ensembl.org) are around 25 000 confidently predicted genes (backed by some form of experimental evidence which is at least a matching EST sequence) and 88 000 predicted by computational methods but with little experimental backing. Other evidence in the literature based on the sensitive method of RT-PCR (reverse transcriptase polymerase chain reaction) suggests that at least some of these predicted but unconfirmed genes will be real, and that the final number of human genes might be 45 000–50 000.

H2 ANNOTATION TOOLS AND RESOURCES

Key Notes

Gene-prediction software	Programs for gene prediction use *ab initio* methods and/or homology searches to identify genes in genomic DNA. Simple programs such as NCBI ORF finder identify ORFs by carrying out six-frame translations. For complex eukaryotic genomes, more sophisticated statistical analysis methods are required. These may incorporate simple rule-based algorithms (e.g. GeneFinder), neural nets (e.g. GRAIL) or Markov models (e.g. Genie, GenScan). Neural nets and Markov model algorithms must be trained on data from specific organisms.
Measuring predictive accuracy	No gene-finding program is 100% accurate and it is better to use several programs to annotate the same genomic sequence. Predictive accuracy is measured in terms of sensitivity (ability to correctly predict genuine genes, or exons) and specificity (ability to correctly eliminate false genes or exons). Common annotation errors include the detection of false coding regions, the merging of exons, failure to detect true exons (especially 5′ and 3′ exons, or very short exons) and the misplacing of intron/exon boundaries. Not all annotation errors are due to limitations of gene prediction software.
Annotation pipelines	The only way to handle the large amount of data from genome projects is to annotate 'on the fly' using a continuous pipeline. Raw sequence data is fed in at one end. This is assembled and verified, annotated using one or more gene-prediction programs and the results are then fed into a database that allows sequence and annotations to be viewed using a graphical interface. Suitable formats include Artemis, Apollo, EnsEMBL, GoldenPath or ACeDB.
Related topics	Sequencing DNA, RNA and proteins (B1) Annotated sequence databases (C2) Genome and organism-specific databases (C3) Sequence similarity searches (E1) Database searches: BLAST and FASTA (E3) Principles of genome annotation (H1)

Gene-prediction software

There are many different programs available for finding genes (*Table 1*). These range from the simple to the extremely complex, and vary according to the gene-finding method used, the basis of the gene-finding algorithm, the range of features identified, and the organisms to which they are applicable. Generally, gene-finding programs can be classed as those using *ab initio* **prediction** methods (identifying genes by looking for general features) and those that also incorporate **homology searches**. The principle features detected by gene-finding programs are discussed in Topic H1.

Table 1. Gene-prediction software available over the Internet

Program	URL
Fgenes, Fgenesh, Fgenesh+	http://genomic.sanger.ac.uk/gf/gfs.shtml
GeneBuilder	http://www.itba.mi.cnr.it/webgene
GeneID	http://www1.imim.es/geneid.html
GeneMarkHMM	http://genemark.biology.gatech.edu/GeneMark
GENIE	http://www.fruitfly.org/seq_tools/genie.html
GeneSCAN	http://genes.mit.edu/GENSCANinfo.html
GlimmerM	http://www.tigr.org/softlab/glimmer/glimmer.html
GRAIL	http://compbio.ornl.gov
HMMGene	http://www.cbs.dtu.dk/services/HMMgene
ORF finder (NCBI)	http://www.ncbi.nlm.nih.gov/gorf/gorf.html.
Wise2	http://www.sanger.ac.uk/Software/Wise2

An example of a very simple gene-finding program is **ORF finder (open reading frame finder)**, which is a component of the National Center for Biotechnology Information (NCBI) suite of programs and can be used over the Internet. This program carries out a six-frame translation of any sequence, and the user is allowed to define the minimum size of the polypeptide as well as the genetic code to be used (Topic H1). The results can be used directly in a BLAST search to identify related sequences in GenBank (Topic E3) or transferred to the Sequin database deposition tool (Topic C2).

As discussed in Topic H1, programs like ORF finder are suitable for the detection of most bacterial genes, which tend to be densely organized and intronless. However, the genes of higher eukaryotes are much more complex, so more sophisticated algorithms are required to detect them. A number of programs are available that detect single features of eukaryotic genes, for example **MZEF** and **HEXON** are **exon-prediction algorithms** that use **hexamer frequency** (the relative frequency with which specific groups of six nucleotides appear in coding vs. noncoding DNA) to discriminate between exons and introns/intergenic sequence. However, most eukaryotic gene-prediction programs can be used to detect and delimit whole genes.

The basis of the algorithm plays an important part in the reliability and scope of gene prediction. Early gene-prediction programs, such as **GeneFinder**, involved **rule-based algorithms** in which an explicit set of rules was used to determine whether a particular nucleotide was included or excluded from a predicted gene. More recently, **neural nets** have been used to combine content and signal sensing. Such programs are capable of learning, and must be trained using annotated sequence data in order to build a set of rules. Programs that use neural nets include **GRAIL (Gene Recognition and Assembly Internet Link)** and **Fgenes.** The difficulties associated with exon-detection in higher eukaryotes stimulated the development of a new range of algorithms based on the statistical analysis of nucleotide frequencies and dependencies, and their integration with signal sensing (Topic H1). These algorithms use **dynamic programming** (Topic F1) and are based on statistical models in which the probability of a given nucleotide appearing in a sequence reflects the preceding nucleotides (this being known as a **Markov model or hidden Markov model)**. Such algorithms are the basis of a number of highly successful gene-prediction programs such as Fgenesh/Fgenesh+, GeneMarkHMM, HMMgene, GeneParser, GlimmerM, Genie, GeneScan and Wise2. Neural net and Markov model-based algorithms can only reliably detect genes in those organisms upon which they have been trained, since they build rules according to experience. For example, GRAIL is suitable for the bacterium *Escherichia coli*, the fruit fly *Drosophila*, the model plant *Arabidopsis thaliana*, mouse and human DNA. GeneScan is suitable for humans (and other vertebrates), *Drosophila*, *Arabidopsis* and maize. Genie is suitable only for *Drosophila* and human sequences, while MZEF can only be applied to human DNA. Recent improvements in such programs allow automatic training on data from new organisms.

Measuring predictive accuracy

With such a large number of gene-finding programs available, how does one know which is the best to use? As discussed above, this may be dictated in part by the source of the genomic sequence, since many of the programs are trained on DNA from particular organisms. However, there is also a useful way to measure the overall accuracy of gene predictions. This is based on the concepts of sensitivity and specificity. Each nucleotide in a given test sequence will either be predicted to be part of a gene (**predicted positive, PP**) or predicted to be outside the boundaries of a gene (**predicted negative, PN**). When a gene is manually annotated, these assignments can be checked against the **actual positive (AP)** and **actual negative (AN)** nucleotides in the sequence. Such comparisons allow four values to be calculated: the number of **true positives (TP), false positives (FP), true negatives (TN)** and **false negatives (FN)** assigned by the program. **Sensitivity** is a measure of the ability of a gene-finding program to detect true positives, and is defined as the ratio TP/AP. Conversely, **specificity** is a measure of the program's ability to eliminate false positives, and is defined as the ratio TP/PP. These values are combined into an overall reliability known as the **approximate correlation (AC)**[1].

Gene prediction software can also be scored on its ability to correctly predict exons. Many annotation errors involve exons, for example failure to detect genuine exons (especially very short exons, or 5′ and 3′ exons since these often contain noncoding sequence), the prediction of false exons, the fusion of exons and the incorrect specification of intron/exon boundaries. A **predicted exon (PE)** may either be a **true exon (TE)** or a **false exon (FE)** and this can be checked against actual **annotated exons (AE)**. **Sensitivity** is defined as the ratio TE/AE and is a measure of the program's ability to correctly predict true exons. **Specificity** is defined as the ratio TE/PE and is a measure of the program's ability to eliminate false exons. Two other values are often used: the **missing exon (ME)** value is defined as the proportion of annotated exons that is not predicted by the software, and the **wrong exon (WE)** value is defined as the proportion of predicted exons that are false.

Gene-prediction programs are continually tested for their ability to correctly identify genes as part of an ongoing international project called **GASP (Gene Annotation aSsessment Project).** At the present time, even the most advanced programs can only achieve about 60% sensitivity and 50% specificity in eukaryote genomes. It is therefore best to use several programs for the analysis of the same sequence. High confidence can be placed in those predictions that correlate across several programs (especially if the algorithms use different statistical methods). Finally, it should be noted that all annotation errors are not the fault of gene-prediction software. Just as many mistakes arise upstream due

[1]For the interested reader, approximate correlation is calculated as follows: $AC=[(TP/(TP+FN))+(TP/(TP+FP))+(TN/(TN+FP))+(TN/(TN+FN))]/2-1$.

to sequencing and sequence assembly errors or due to misleading information already present in the databases!

Annotation pipelines

The large amount of sequence data generated by genome projects means that annotation must be carried out continuously. The data are fed in at one end and the annotations flow from the other. This is the nature of an **annotation pipeline**. The pipeline begins with the sequencing work, assembly and quality control (Topic B1). The finished data are then mined for information using the prediction tools discussed in this section and then integrated into an appropriate database. Database formats suitable for the analysis and display of annotated genome data are discussed in Topic C3. These include Artemis, Apollo, EnsEMBL, GoldenPath and ACeDB.

I1 CONCEPTUAL MODELS OF PROTEIN STRUCTURE

Key Notes

Structural types and conceptual models	Globular proteins are soluble in predominantly aqueous solvents such as the cytosol and extra-cellular fluids, and integral membrane proteins exist within the lipid-dominated environment of biological membranes. Conceptual models of protein structure are valuable aids to understanding protein bioinformatics.
Globular proteins	In globular proteins, the linear amino acid polymer forms a three-dimensional structure by folding into a globular compact shape. Globular proteins tend to be soluble in aqueous solvents and folding is dominated by the hydrophobic effect, which directs hydrophobic amino acid side-chains to the structural core of the protein, away from the solvent.
Secondary structure	Globular proteins usually contain elements of regular secondary structure, including α-helices and β-strands. These are stabilized by hydrogen bonding and contribute most of the amino acids to globular protein cores. Residues in regular secondary structures are given the symbols H, meaning helix, or E (or B), meaning extended or β strand.
Tertiary structure	The tertiary structure is the full three-dimensional atomic structure of a single peptide chain. It can be viewed as the packing together of secondary structure elements, which are connected by irregular loops that lie predominantly on the protein surface. Loop residues are given the symbol C to distinguish them from residues in helices or strands.
Quaternary structure	Several tertiary structures may pack together to form the biologically functional quaternary structure.
Integral membrane proteins	These exist within biological lipid membranes and obey different structural principles compared with globular proteins. They contain runs of generally hydrophobic amino acids, associated with membrane-spanning segments (often but not exclusively helices), connected by more hydrophilic loops that lie in aqueous environments outside the membrane. Membrane proteins are very important components of cellular signaling and transport systems.
Domains	Proteins tend to have modular architecture and many proteins contain a number of domains, often with mixed types, for example mixed integral membrane and globular domains.
Evolution	In globular proteins, surface residues in loops evolve (change) more quickly than residues in the hydrophobic core. In integral membrane proteins, the most slowly evolving residues are those in the membrane-spanning regions.

Related topics	Amino acid substitution	The relationship of protein
	matrices (E2)	three-dimensional structure
	Multiple sequence alignment and	to protein function (I2)
	family relationships (F1)	The evolution of protein structure
		and function (I3)

Structural types and conceptual models

Conceptual models can add enormously to our understanding of protein sequence, structure and function. It is the purpose of this topic to introduce some key conceptual ideas that are essential to understand protein bioinformatics. It is useful to distinguish three different protein structural types: **fibrous proteins** (e.g. collagen), **globular proteins**, which tend to exist in aqueous solvents like the cytosol and extracellular fluids, and **integral membrane proteins**, which exist within the lipid environment of biological membranes. In this text we will be concerned mainly with globular proteins and integral membrane proteins.

Globular proteins

Proteins with a wide variety of functions fall within this broad class, including enzymes, antibodies, and a variety of molecules associated with signaling and transport. Overall they tend to have globular (spheroidal) shapes. Structure formation in globular proteins is dominated by the **hydrophobic effect**. Nonpolar chemical compounds with few charged atoms are **hydrophobic**: they do not dissolve easily in water. In contrast, polar compounds with charged atoms form electrostatic and hydrogen bonding interactions with water molecules and therefore tend to dissolve easily. Some of the amino acid side-chains are hydrophobic (valine, leucine, etc.), while other amino acids are hydrophilic (aspartic acid, lysine, etc.). When a protein folds, it is able to minimize its free energy by placing hydrophilic amino acids on the surface of the globule, in contact with the aqueous solvent, and hydrophobic amino acids in the central core of the globule, away from the solvent. It is thought that this effect is the strongest force driving the linear polymer of amino acid residues to fold into globular shape in water. Although most core side-chains are hydrophobic, there are sometimes some hydrophilic side-chains in the core, very often making hydrogen bonds or ionic interactions with other core side-chains, and equally, there is often a small proportion of hydrophobic side-chains on the protein surface. An important feature of protein cores is that they are efficiently packed, so that most of the available space is filled with atoms.

Secondary structure

Globular proteins usually contain elements of regular **secondary structure**. The best-known example is the **α-helix**, where four or more consecutive amino acid residues in the polymer adopt the same conformation, resulting in the adoption of a regular helical shape by the polypeptide backbone. This helix is stabilized (held together) by hydrogen bonds between the main chain C=O group of each amino acid and the H–N group of the amino acid four residues further along the helix. This forms a helix with 3.6 amino acid residues per helical turn (see *Fig. 1*). Although helices with slightly different hydrogen bonding patterns (3_{10} and π helices) do occur in protein structures, α helices are by far the most common. The illustration in *Fig. 1* shows that the amino acid side-chains point away from the helical axis, forming a surface for the helix. The notation H is used to indicate that a particular residue is a member of an α-helix.

(a)

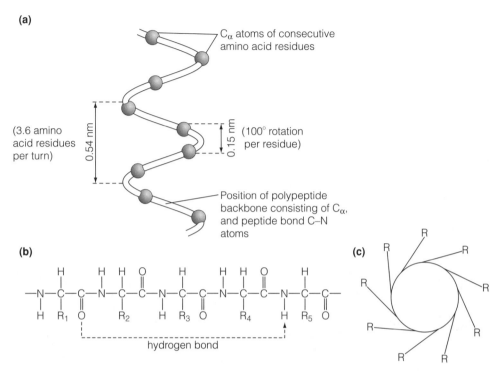

Fig. 1. The folding of the polypeptide chain into an α helix. (a) Model of an α helix with only the C_α atoms along the backbone shown; (b) in the α helix the CO group of residue n is hydrogen bonded to the NH group on residue (n+4); (c) cross-sectional view of an α helix showing the positions of the side-chains (R groups) of the amino acids on the outside of the helix. From Hames B.D. and Hooper N.M., Instant Notes in Biochemistry, 2nd edition, © BIOS Scientific Publishers Limited, 2000.

The other common type of regular secondary structure is the **β-strand** (see *Fig. 2*). This secondary structure element is formed by consecutive amino acids in their most extended conformation, and for this reason the letter E (Extended) is used to indicate that a residue adopts this structure. In this case, hydrogen bonds between main chain C=O and N–H groups form not to residues in the same strand, but to residues in strands formed by other parts of the polypeptide. These hydrogen bonds mean that single beta strands do not exist in isolation, but are always spatially adjacent to at least one other strand. The twisted, pleated structure formed by consecutive, spatially adjacent, hydrogen-bonded strands is known as **β-pleated** sheet. If such a sheet is curved round so that the stands that would have been on the edges of the sheet are spatially adjacent and hydrogen bonded, the structure is known as a β barrel. The illustration in *Fig. 2* shows that the amino acid side-chains point alternately above and below the sheet.

Tertiary structure   The **tertiary structure** or **fold** of the protein is the position of every protein atom in three-dimensional space. This can be considered conceptually as the product of a process in which the secondary structure elements of the polypeptide chain pack together with their inward facing sides contributing most of the residues to the hydrophobic core. These secondary structure elements are connected by sections of polypeptide chain called **loops**. For our purposes, we can consider

(a)

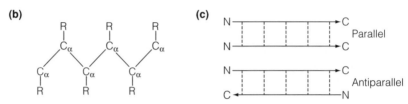

(b)   **(c)**

Fig. 2. The folding of the polypeptide chain in a β-pleated sheet. (a) Hydrogen bonding between two sections of a polypeptide chain forming a β-pleated sheet; (b) a side-view of one of the polypeptide chains in a β-pleated sheet showing the side-chains (R groups) attached to the C_α atoms protruding above and below the sheet; (c) because the polypeptide chain has polarity, either parallel or antiparallel β-pleated sheets can form. From Hames B.D. and Hooper N.M., Instant Notes in Biochemistry, 2nd edition, © BIOS Scientific Publishers Limited, 2000.

the loops to be of irregular structure (sometimes called **coil** or **random coil**, and denoted by the letter C), although a more detailed treatment would reveal some regular structures. The loops tend to lie on the protein surface and to contain mainly hydrophilic residues.

Quarternary structure

Many proteins exist as multimeric molecules formed as several polypeptide chains bind together to form a complex. This complex is known as the quaternary structure of the protein, and it is the biologically functional unit. Sometimes complexes have a single biochemical function, but there are many examples of multifunctional complexes, for instance enzymes that catalyze consecutive steps in a metabolic pathway.

Integral membrane proteins

Integral membrane proteins are key elements of biological signaling and transmembrane transport systems. Examples are the G protein-coupled receptors (involved in signaling), and channel proteins responsible for rapid transport of ions across membranes. The fact that significant parts of these proteins exist in the lipid environment of biological membranes, in contrast to globular proteins that exist in aqueous solvents, means that different structural principles apply to them. Typically, integral membrane proteins have one or more segments that actually cross the membrane. Amino acids in these segments tend to be of hydrophobic nature, compatible with the lipid structure of the membrane, while portions of the protein that exist in the more aqueous environment on either side of the membrane tend to be more polar. Very often the membrane-spanning segments adopt an α-helical structure, as in G protein-coupled receptors and many ion channels, but there are examples, for instance the bacterial porins, where the membrane-spanning segments are β strands. A schematic illustration of a membrane protein is shown in *Fig. 3*. The number of membrane-spanning sections within a typical membrane protein varies from one (often a membrane 'anchor' for the protein), to more than 10. The well-known G protein-coupled receptors, like rhodopsin, contain seven membrane-spanning segments.

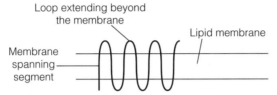

Fig. 3. An integral membrane domain with seven membrane-spanning segments.

Domains

From the knowledge of protein sequence and structure we have now, it is clear that many proteins have a modular architecture. Nature creates proteins with complex functions by combination of quasi-independent modular units or **domains**, typically with much simpler functions. For instance there are many eukaryotic proteins that contain the homeo domain whose function is to bind to DNA, but these proteins also contain other domains that are responsible for other aspects of their overall function. This is illustrated in *Fig. 4*. The precise definition of a protein domain depends on your point of view. Some would define it as a protein sub-sequence or substructure with a recognized function, as above. Others view domains as protein sub-sequences that are able to fold independently of the rest of the protein, and others regard domains as geometrically distinct parts of the protein structure. Perhaps the most useful definition of a domain, however, is that it is a protein unit observed to occur in many otherwise unrelated proteins.

Whichever your preferred definition of a protein domain, it is very important to bear in mind the modular architecture of proteins in many bioinformatics analyses. It is also important to appreciate that proteins can have domains of mixed types, for instance integral membrane domains are often found in combination with globular domains. This is the case for many receptors, which are anchored to a biological membrane by an integral membrane domain, and have extracellular globular domains responsible for recognizing biochemical signals.

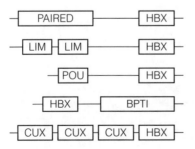

Fig. 4. Multidomain proteins containing a homeo domain. The sequences are shown from amino terminus (left) to carboxy terminus (right) with recognized domains shown as boxes. From top to bottom the proteins are human PAX4, human IHX2, human OTF-2, C. elegans c02f12.5 and mouse Cux-2. HBX, homeo domain; PAIRED, paired domain; LIM, LIM domain; POU, POU domain; BPTI, Kunitz BPTI domain; CUX, CUX domain.

Evolution   In earlier sections, we explained that changes in the amino acid sequences of proteins occurring during evolution are accepted if they have a neutral effect on protein structure and function (i.e. they have little effect on the stability of the protein and its ability to perform its normal function), or if they have a positive effect (i.e. they enhance stability or functional efficacy). When a change has a negative effect it is likely to be rejected by natural selection, except in the cases of the changes occurring when a duplicated gene is in the process of evolving a new function.

When we observe the evolution of protein structures we see that evolutionary change is very slow for residues in the structural core, and significantly faster for residues on the protein surface. This can be understood in light of the conceptual models introduced above. Within the structural core the introduction of changes can affect the tight atomic packing, resulting in reduction of protein stability, and is therefore less likely. On the other hand, residues on the protein surface are not subject to this constraint (essentially they need only to be hydrophilic) and are therefore much more susceptible to evolutionary change. Because secondary structure elements tend to contribute most of the core residues, while loop residues lie mainly on the surface, it is often observed that loops evolve much more quickly than secondary structure elements. A related effect is observed in integral membrane proteins, where the membrane-spanning segments (most often helical) evolve more slowly than the loops that connect them.

I2 THE RELATIONSHIP OF PROTEIN THREE-DIMENSIONAL STRUCTURE TO PROTEIN FUNCTION

Key Note

Structure and function	Proteins rely upon the shapes and properties of key functional areas of their three-dimensional structures to carry out biological functions. Knowledge of protein structure is key to understanding protein function and this is one reason for its importance in bioinformatics.

Related topics Conceptual models of protein structures (I1) The evolution of protein structure and function (I3)

Structure and function

The ability of proteins to perform biological functions depends on the formation of a three-dimensional structure (fold) that is stable in the normal environment of the protein. For example, enzymes often catalyze reactions using an **active site**. Typically, this is a cavity in the three-dimensional structure of the enzyme that is accessible to the reactants from the protein surface. Active sites are multifunctional. They contain the key catalytic machinery of the protein, which is typically one or more residues that are actively involved in the chemical reaction catalyzed, and stabilize transition states. They exclude solvent from the reaction, and their shape and physicochemical properties are such that they bind the intended reactants much more strongly than any alternatives, thereby creating catalysis specific to certain molecules or molecular classes. All this depends on the active site adopting a specific three-dimensional shape, and ultimately on the fold of the peptide chain itself. The importance of the three-dimensional form of the active site in serine protease enzymes is illustrated in *Fig. 1*.

Not all proteins are enzymes, but most rely on their three-dimensional structures in order to perform biological functions. Molecular recognition without catalysis is a function of many proteins. Transport proteins need to recognize the molecules they carry. Protein–protein recognition is important in the recognition of foreign proteins by antibodies, the interactions between components of signaling pathways, and the formation of multifunctional complexes. Similarly, the recognition of other macromolecules by proteins is important in the regulation of gene expression (by DNA binding proteins), and the formation of mixed protein–RNA complexes like the ribosome. Finally many receptor proteins need to specifically recognize a molecular signal (for example, the recognition of steroid hormones by receptors in the cell nucleus).

For molecular recognition to take place, it is necessary that the molecules concerned are able to bind together in an energetically favorable conformation. This depends on **complementarity**. They must be able to form complementary

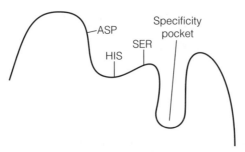

Fig. 1. A schematic view of a serine protease active site where the line represents the protein surface. The key catalytic triad comprising residues serine (SER), histidine (HIS) and aspartic acid (ASP) is shown. During catalysis, the serine residue is activated as a nucleophile by transfer of charge to the other key residues in order to attack and cleave peptide bonds. Also shown is the specificity pocket responsible for recognizing specific types of molecules for catalysis. While all serine proteases share the catalytic triad, the shape and properties of the specificity pocket vary to produce enzymes of different specificities. In trypsin, the pocket recognizes peptides with basic side-chains, while in chymotrypsin, it recognizes large hydrophobic side-chains. The shape and properties of the active site are created by residues from distal parts of the sequence, which are brought close together in space by the three-dimensional fold of the peptide chain.

shapes, so that they fit snugly together, and where their surface atoms are in contact there must be complementarity of physicochemical properties. This means that negatively charged areas of one interacting partner must contact positively charged areas of the other, and hydrophobic areas must be in contact with other hydrophobic areas. All this depends on the formation of a stable three-dimensional structure by the protein.

I3 THE EVOLUTION OF PROTEIN STRUCTURE AND FUNCTION

Key Notes

Structural and functional constraints	Evolution accepts changes to amino acid residues in proteins where they have a neutral or advantageous effect on protein structural stability or protein function. Residues can be conserved for structural or functional reasons. Amino acids are conserved where they are uniquely able to fulfill particular structural roles. This often occurs with cysteine, glycine and proline.
Multiple sequence alignment	Understanding how structures evolve can help us understand multiple sequence alignments. Key structural and functional residues are often observed to be conserved. Insertions and deletions are seen to occur preferentially in hydrophilic surface loops by comparison with regular secondary structure elements. Loops are also subject to faster mutational change. Conservation of hydrophobic core residues in secondary structure elements is also common, as are conservation patterns associated with amphipathic helices.
Evolution of the overall protein fold	If two naturally occurring protein sequences can be aligned to show more than 25% sequence similarity over an alignment of 80 or more residues, then they will share the same basic structure. The Sander–Schneider formula gives the higher threshold percentage identities necessary to guarantee structural similarity from shorter alignments.
Conservation of structure	Protein structures tend to be conserved even when evolution has changed the sequence almost beyond recognition. Structural knowledge is therefore a key factor in understanding protein evolution.
Evolution of function	While structure tends to be conserved by evolution, function is observed to change. There are many examples of proteins whose sequence and structure are very similar, but which have different functions. When function has changed, key functional residues change as well, and this is often clear in multiple sequence alignments.
Related topics	Amino acid substitutions matrices (E2) Multiple sequence alignment and family relationships (F1)

Structural and functional constraints

The evolution of protein sequences was discussed in Topic E2 where amino acid substitutions were understood in terms of physicochemical relationships between the different amino acids. Amino acids with similar size and physico-chemical properties are likely to make reasonable replacements for each other, and to form accepted substitutions during protein evolution. With structural knowledge, we can be more precise. Fundamentally, amino acid substitutions are accepted in evolution if the change is either neutral or advantageous to protein structural stability and/or function. Although matrices such as PAM250 (Topic E2) can tell us about average rates of substitution between amino acids, when the three-dimensional structure is known we are able to understand which substitutions are acceptable at particular positions in the protein sequence.

Some amino acid residues have key roles to play in the stability of particular structures (see *Table 1*). When a residue plays a key role it is often the case that no other residue can substitute for it while maintaining the stability of the structure, and the residue is conserved throughout the evolutionary family. In a similar way, key functional residues are also often conserved. Many enzymes rely on the chemical properties of certain amino acids to effect catalysis, for instance serine, histidine and aspartic acid residues in the serine protease active site (Topic I2). If these properties are unique to the residue involved, it will tend to be conserved to preserve function.

Multiple sequence alignment

The relationship between multiple sequence alignments (Topic F1) and protein structure and function is very important. Many of the features of the alignment can be understood in the light of a known structure for one member of the family. *Figure 1* shows a multiple sequence alignment of lysozyme and α-lactalbumin sequences with key conserved structural (disulfide-forming cysteines) and functional (key catalytic residues) features annotated.

In Topic I1, we explained that residues on the protein surface evolve more quickly than residues buried in the structural core, because they are not subject to the strong constraints involved in the maintenance of efficient packing of the structural core. We also described a conceptual model of globular protein structure in which secondary structure elements (helices and strands) contribute most residues to the hydrophobic core of the protein and alternate with loops whose residues lie predominantly on the surface. Loops are therefore observed to evolve more quickly than secondary structure elements. This effect is

Table 1. Unique structural roles played by some amino acid residues in protein structures. When these roles are adopted the residue concerned is often found to be conserved throughout the protein evolutionary family, because loss of the role, which no other residue could fulfill, would incur a loss in structural stability

Residue	Structural role
Cysteine	The formation of disulfide bridges (S–S bonds) with other cysteine residues, making covalent links between sequence separated parts of the polypeptide chain. These bonds can make a major contribution to structural stability.
Glycine	The side chain of glycine is very small (a single hydrogen atom). This means that the structure of glycine is more flexible than that of the other amino acids. It can adopt some conformations that are accessible to no other residues, and is often found where the peptide chain goes through a very tight turn.
Proline	Amino acids tend to prefer to adopt *trans* conformations around the planar peptide bond, that is, with the two α carbons located on opposite sides of the bond. *Cis* conformations (with the α carbons on the same side of the bond) occur occasionally, and often involve a proline residue, which is better able to form the *cis* conformation than the other amino acids.

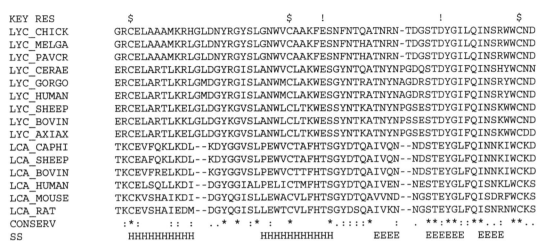

Fig. 1. A multiple alignment of lysozyme and α-lactalbumin sequences. Sequence names starting LYC are lysozymes and those starting with LCA are α-lactalbumins. The KEY RES lines indicates key structural residues (disulfide-forming cysteines) with the symbol '$', and key lysozyme catalytic residues with the symbol '!'. Key structural residues are conserved in all sequences, but the lysozyme functional residues are not conserved in the α-lactalbumin sequences. The CONSERV line indicates the degree of conservation in a particular column ('*'=identically conserved, ':' contains only very conservative substitutions, '.' contains conservative substitutions). The SS line shows secondary structure (taken from one sequence of known 3D structure).

observed in multiple sequence alignments, where it is often clear that the alignment consists of alternating blocks of well-conserved residues corresponding to secondary structure elements and less well-conserved residues from surface loops. This is evident in the alignment shown in *Fig. 1*, where it is clear that insertions and deletions in the sequences fall in loops (outside the labeled secondary structure elements), and that strong conservation of hydrophobic residues is observed in some secondary structure elements. For example, the second helix contains a group of three residues of hydrophobic character, which are W, V and C in the first (top) sequence.

As we commented in Topic F1, multiple alignments made on the basis of sequence alone by most standard software methods are often not perfect. Structural information can be used to make manual improvements. For instance, if it is known that a particular cysteine is involved in a disulfide bridge it should be verified that it is aligned with a cysteine in all sequences in the family. Similarly, if the alignment involves insertions and deletions, it should be borne in mind that these are much more likely in loops than in secondary structure elements.

In the case where there is no known three-dimensional structure, a multiple alignment can serve as a first stage in structure prediction. Conserved cysteines can be predicted to be involved in disulfide bonds (although this is not the only reason for conservation of this residue, which is also often used to bind a metal ligand). Similarly, the alternation between blocks of conserved residues and blocks where the sequences are more variable, which is apparent in many multiple alignments, can be interpreted in terms of probable secondary structure elements alternating with surface loops.

Another structural feature that is often visible in multiple sequence alignments is the amphipathic helix. In Topic I1 we showed that the amino acid side-chains point away from the helical axis and form a surface for the helix. Very often, α helices are positioned in protein structures so that one side of the helix is part of the hydrophobic structural core and the other side lies on the protein surface. Hydrophobic side-chains therefore dominate on one side and hydrophilic ones on the other. Since an α helix has 3.6 residues per turn, an amphipathic helix tends to exhibit an alternation between hydrophilic and hydrophobic residues with a periodicity of three to four residues. This is often visible in multiple sequence alignments as conserved hydrophobic residues occurring alternately every three or four residues in the sequence. This is illustrated in *Fig. 2*. Helical wheel illustrations of the type shown in *Fig. 2* are produced by various software programs available from the text WWW site.

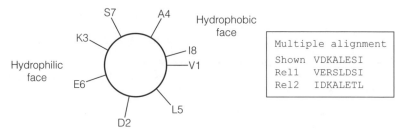

Fig. 2. An amphipathic α helix. The helical wheel (left) shows the positions of the side-chains for two turns of the α helix. The helical wheel is a projection of the helix down its own axis onto a plane, showing the positions of the side chain. For instance, in this case residue I8 is about two helical turns from V1, on the same side of the helix but almost directly above). The helix has a hydrophobic face populated by A4, I8, V1 and L5 and a hydrophilic one populated by D2, E6, K3 and S7. The multiple alignment of the sequence shown and two relatives illustrates the conservation of hydrophobic residues alternately four (V1 and L5) and three (L5 and I8) residues apart in the sequence.

Evolution of the overall protein fold

So far, we have considered the evolution of parts of the protein structure in isolation, now we consider how the overall structure evolves. In 1991, Sander and Schneider published a study of naturally occurring proteins of known structure showing how much similarity in protein sequence is needed to guarantee similarity in structure. They found that the threshold of similarity depended on the lengths of the sequences involved, with more similarity being required for shorter sequences. They showed that if two protein sequences can be aligned to show more than 25% identical residues over an alignment length of more than 80 residues, then they will share the same basic three-dimensional structure. This result only applies to the naturally occurring proteins currently of known structure. There are now examples of artificially engineered proteins with sequence similarity of as much as 50% yet with different structures. Other potential exceptions to Sander and Schneider's rule are proteins like the prion protein that appear to have two distinct possible structures (one normal and the other disease associated). The Sander–Schneider rule was derived from a database in which the proteins were in their normal biologically active conformations, and these are the conformations to which it applies.

Sander and Schneider's threshold for structural similarity was

$$t\,(L) = 290.15L^{-0.562}$$

where L is the length of the alignment and t is the percentage identity threshold required to guarantee structural similarity. For alignments with L around and greater than 80 the value of t approaches the 25% mentioned above, but it is larger for smaller values of L. For instance, a similarity of $t = 43\%$ is required for structural similarity in an alignment of $L = 30$ residues.

Sander and Schneider's result was very important because it provided the theoretical basis for a method of structure prediction known as comparative modeling (see Topic I8 for more details). Using the fact that sequences with a sufficiently high level of similarity can be assumed to have the same structure, sequences of known structure can be used in structure predictions for sufficiently similar sequences of unknown structure.

Conservation of structure

Sequences with sufficient similarity share the same structure, but observations made on known sequences and structures have shown us that divergent evolution of proteins with a common ancestor can operate until their sequence similarity is almost undetectable. The best-known example of this comes from the globin family, including the hemoglobin chains, myoglobin, and plant leghemoglobin. These sequences adopt the same overall structure, and carry out similar functions (oxygen transport) by the same mechanism, but often exhibit sequence identities that are below 20% of identical residues. This is a level of sequence similarity that is often found between the sequences of completely unrelated proteins. This observation that sequence evolves much faster than structure means that structure is very important to bioinformatics. Many potential evolutionary relationships that were not apparent on the basis of sequence analysis have been discovered when three-dimensional structures have become available.

Evolution of function

We have commented in previous sections that protein function evolves, so that proteins that are clearly homologous (i.e. have closely related sequences and/or structures) sometimes have different functions, and we have identified gene duplication as a possible mechanism for the development of new functions. When this happens, residues that have a key role in the performance of one function that are conserved throughout the members of the family performing that function are often not the same or even conserved in members of the family which perform a different function. Lysozyme and α-lactalbumin are examples of homologous proteins that perform very different functions (the former is an enzyme and latter a regulatory protein). The multiple alignment in *Fig. 1* shows that the key lysozyme functional residues are not conserved in α-lactalbumin, while key structural residues are (in particular the disulfide bond-forming cysteine residues).

Gene duplication is not the only way in which protein function evolves. Multidomain proteins (see Topic I1) are proteins that consist of more than one domain. Often the different domains have different functions that are combined to produce another, much more complex function. For example, some enzymes have domains associated with regulation of their activity. It seems that nature evolves many new proteins by swapping and recombination of modular units (domains) and that this is a major route for the evolution of new and more complex protein functions.

14 OBTAINING, VIEWING AND ANALYZING STRUCTURAL DATA

Key Notes

Software and WWW sites	A large variety of software for structure visualization and analysis is available on the WWW.
Obtaining data	All published protein structures are submitted to a public database. Data search and download can be performed at various WWW sites.
Visualization of structures	Rasmol, Chime and Cn3d are commonly used programs for viewing structural data.
General structural analysis	There is an enormous amount of software available for structural data analysis, and also several WWW sites holding pre-prepared analyses.
Analysis of functional sites	Functional sites in protein structures typically contain a few residues in defined spatial positions. Software and databases have been developed to locate and search for similarity in such sites.

Related topics	Determination of protein structure (B2)	Data retrieval with Entrez and DBGET/Link DB (D1)
	Miscellaneous databases (C4)	Data retrieval with SRS (D2)
		Cheminformatics resources (M2)

Software and WWW sites

It would be impossible to produce a comprehensive list of available software and World Wide Web (WWW) sites for the display and analysis of protein and other macromolecular structures. All we provide here are pointers to the most commonly used and freely available utilities. The software and WWW sites we discuss are all linked from the text WWW site. It should be noted that this book is not intended to serve as a software manual for commonly used programs; that function is much better provided by the documentation that is distributed with the software itself. Such documentation will be much more extensive than the space available here would allow, and should remain up to date with changes to programs as they are made.

Obtaining data

We have already discussed the structural database (Topic C4) and the experimental methods that produce the data contained within it (Topic B2). Like the sequence databases discussed in earlier chapters, the structural database can be searched by many of the standard search tools, including SRS (Topic D2) and tools available at the National Center for Biotechnology Information (NCBI) WWW site (Topic D1). There are also specialized search engines maintained by the macromolecular structural database group at the European Bioinformatics Institute (EBI; http://msd.ebi.ac.uk), and the Research Collaboratory for Structural Bioinformatics (RSCB; http://www.rcsb.org/pdb). Both these tools allow easy downloading of structural data in the standard Protein Data Bank (PDB) format, which is input to much of the software discussed below.

Visualization of structures

The program **RasMol** (*Table 1*), written by Roger Sayle, is perhaps the best-known viewer for macromolecular structures. Given a structure in the standard structural database format (Topic C1), this software displays a three-dimensional image of the structure. The image can be rotated by use of the mouse to produce different views, and displayed in various formats, including **wireframe** (bonds displayed as lines with atoms implied at junctions and ends), **space filling** (each atom represented by sphere of proportionate size), and **ball and stick** (atoms represented by small spheres and chemical bonds by sticks). There are also special **cartoon** formats that give clear displays of secondary structure elements. The user can choose between various color schemes and even use customized colors. Finally, there are flexible ways of selecting parts of structures to enable highlighting with a different display format. For instance, key residues or substructures can be highlighted by variation in the display mode. *Figure 1* shows a small protein structure displayed in RasMol in two different display formats, the first showing all atoms, and the second abstracted to show only secondary structure elements, with helices as helical ribbons and stands as extended ribbons. A more complex example of an enzyme with its inhibitor displayed in different formats is shown in *Fig. 2*, along with the RasMol commands that produced the display.

Some other related programs are Chime (a plug-in for use in a WWW browser like Netscape), which has similar functionality to RasMol, and Cn3d, which is also able to link the structure display directly to a multiple sequence

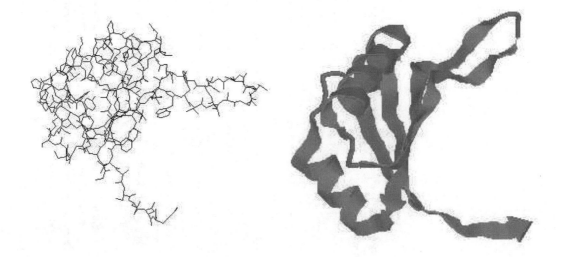

Fig. 1. Two images of protein structures displayed with the RasMol program. On the left all atoms are displayed in a format known as 'wireframe'. The image on the right displays the secondary structure elements of the protein fold in a format known as 'cartoon'. The fold on the right comprises a four-stranded antiparallel β sheet packed against two α helices.

Table 1. Summary of software and WWW sites for protein structure visualization and analysis (links are all on the text WWW site)

Resource	Type	URL	Function
RasMol	Computer program for most computer operating systems	ftp://ftp.dcs.ed.ac.uk/pub/rasmol/ http://www.bio.cam.ac.uk/doc/rasmol.html	Visualization of protein structures in three dimensions
Cn3D	Helper application to permit viewing of 3D structures in a WWW browser, for most computer operating systems	http://www.ncbi.nlm.nih.gov/Structure/CN3D/cn3d.shtml	Visualization of protein structures in three dimensions, which can be linked to sequence alignments
Chime	Helper application to permit viewing of chemical structures in a WWW browser	http://www.mdli.com/	Visualization of protein structures in three dimensions
Molscript	Computer program for UNIX operating systems	http://www.avatar.se/molscript/	Visualization of protein structures in three dimensions
Ribbons	Computer program for most operating systems	http://sgce.cbse.uab.edu/ribbons/	Visualization of protein structures in three dimensions
TOPS	Computer program for UNIX-related operating systems and WWW server	http://www.tops.leeds.ac.uk	Program for visualization of protein folding topologies
DSSP	Computer program for most computer operating systems	http://www.cmbi.kun.nl/gv/dssp/index.html	Finds secondary structure elements in an input protein structure. Also calculates relative solvent accessibility
MSMS	Computer program for UNIX-related operating systems	http://www.scripps.edu/pub/olson-web/people/sanner/html/msms_home.html	Protein surface calculation
Surfnet	Computer program for UNIX-related operating systems	http://www.biochem.ucl.ac.uk/bsm/biocomp/index.html#software	Visualization of protein surfaces
HBPLUS	Computer program for UNIX-related operating systems	http://www.biochem.ucl.ac.uk/bsm/biocomp/index.html#software	Finds internal hydrogen bonds and nonbonded interactions.

NACCESS	Computer program for UNIX-related operating systems	http://www.biochem.ucl.ac.uk/bsm/biocomp/index.html#software	Calculates atomic and residue solvent accessibilities
PROCHECK	Computer program for UNIX-related operating systems	http://www.biochem.ucl.ac.uk/bsm/biocomp/index.html#software	Checks stereochemical quality of protein structures
PROMOTIF	Computer program for UNIX-related operating systems	http://www.biochem.ucl.ac.uk/bsm/biocomp/index.html#software	Analyses protein structural motifs
LIGPLOT	Computer program for UNIX-related operating systems	http://www.biochem.ucl.ac.uk/bsm/biocomp/index.html#software	Produces graphical displays of ligands and their binding sites
PDBSum	WWW database	http://www.biochem.ucl.ac.uk/bsm/pdbsum/	A WWW resource containing detailed summaries of structural analyses carried out on entries in the public structural database
PROCAT	WWW database	http://www.biochem.ucl.ac.uk/bsm/PROCAT/PROCAT.html	A database of three-dimensional atomic templates defining particular enzymatic activities, taken from active sites in known enzyme structures
Relibase	WWW database and search server	http://relibase.ebi.ac.uk	A database of protein ligand complexes with geometrical search facilities
SPASM	WWW search server	http://portray.bmc.uu.se/cgi-bin/dennis/spasm.pl	Geometrical searches for structural motifs
ASSAM	Computer program	–	Geometrical searches for structural motifs

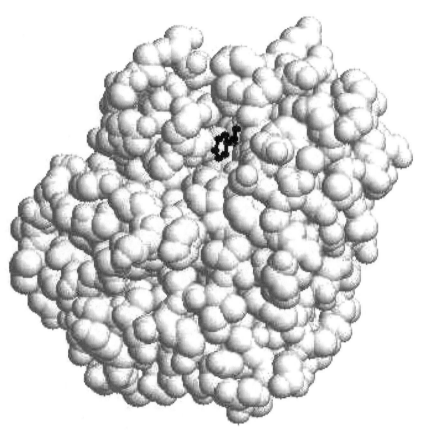

Fig. 2. The structure of the serine protease enzyme trypsin in 'space-filling' format with an inhibitor (benzamidine) shown in black, bound in the main specificity pocket of the enzyme, adjacent to the active site.

alignment. The visualization of protein structures in three dimensions can be very difficult, and for this reason there are several tools that produce schematic or summary displays of protein folds. One example of this is the RasMol cartoons format (*Fig. 1*), and similar more sophisticated displays can be produced with Molscript and Ribbons. Finally, TOPS is a utility that produces two-dimensional summaries of three-dimensional protein folds, an example of which is shown in *Fig. 3*.

General structural analysis

Protein structures offer almost unlimited possibility for analysis programs, including the automatic annotation of secondary structure elements, the analysis of residue solvent accessibility to find surface and core residues, the creation protein surfaces, checking the stereochemical quality of structures, and analysis and display of functional sites. Some examples of such programs are listed in *Table 1*. One WWW resource that is worthy of special mention is PDBSum, provided by the bioinformatics group at University College London. This resource contains a well-presented summary of structural data and analyses for every entry in the structural database, accessed by the accession code for that database (Topic C4). The information presented includes secondary structure, disulfide bonds, ligand binding sites, active sites, key residues, plots

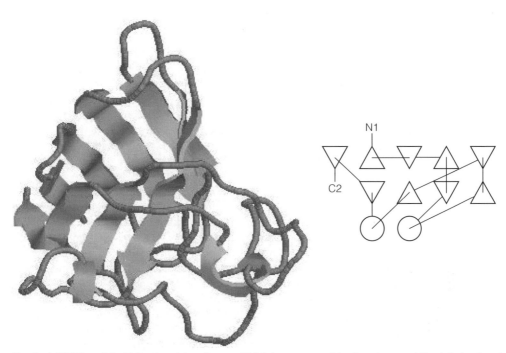

Fig. 3. A TOPS protein-folding topology diagram (right) for a superoxide dismutase and the full structure in RasMol cartoons format (left). The RasMol diagram shows beta strands as arrows and alpha or 3_{10} helices as helical ribbons. The core of the protein comprises two antiparallel β sheets, one having four strands and the other five. The sheets pack face-to-face to form a sandwich structure, and two short helices lie in loops in front of the four-stranded sheet. This is difficult to see from the three-dimensional RasMol diagram but is much clearer in the TOPS cartoon. In the TOPS cartoon, the peptide chain runs from N-terminus N1 to C-terminus C2 and can be traced by following the connecting lines from symbol to symbol. The triangular symbols represent beta strands and the circular ones helices. The symbols should be thought of as representing secondary structure elements, which are perpendicular to the plane of the diagram. They have a direction (N to C), which is either 'up' (out of the plane of the diagram) or 'down' (into the plane of the diagram). 'Up' strands are represented by upward-pointing triangles and 'down' ones by downward-pointing triangles. The five-stranded sheet is represented by the upper horizontal row of five triangles and the four-stranded sheet by the lower one. The alternation of strand directions along these rows shows that the sheets are antiparallel. The observant reader may have noticed that the connecting lines between the triangular strand symbols are drawn sometimes to the edge of the symbol and sometimes to the center. This is in fact another way of showing the direction of the secondary structure element: if the N-terminal connection is drawn to the edge of the symbol and the C-terminal connection to the center, then the direction is up; otherwise the N-terminal connection is drawn to the center and the C-terminal one to the edge, and the direction is down. So there are two ways in which the direction of strands is shown, but the direction of helices must be deduced by looking at the connecting lines.

of intermolecular interactions, folding topologies, EC numbers for enzymes, and much more.

Analysis of functional sites

Probably the most important parts of any protein structure are its functional sites, that is, active sites of enzymes, ligand binding sites, metal ion binding sites, sites where other proteins or DNA might bind, etc. Location of such sites in the database of known structures is surprisingly difficult. Some sites are annotated by those who determined the structures, and others can be located because they contain bound ligands, and most of these are part of PDBSum (described above). Nevertheless, many potentially interesting sites in known protein structures cannot be found in this way, and there is a need for methods that can locate such sites as part of the functional analysis of new protein structures.

To analyze a new structure it is necessary to first locate potential functional sites and then consider what the structure of these sites can reveal about possible functions. It has been observed that enzyme active sites are often located in large cavities near the protein surface (more often than not the active site is the largest such cavity). The SURFNET software (*Table 1*) is able to locate such sites. To make a prediction about the actual function of any potential site on a protein surface is a problem currently at the forefront of bioinformatics research. In common with much of bioinformatics, most current approaches are knowledge based, and rely heavily on databases of information about active sites of known function. This is based on the idea that active sites that are in some sense similar are likely to have similar functions.

There are several databases, computer programs and resources that implement similarity searches for functional sites in proteins. Such similarity searches are both geometrical and chemical: they search for sites where the relative geometrical positions and chemical type of atoms, residues or chemical groups are similar. PROCAT is a manually curated database of three-dimensional atomic active site templates, which define particular enzymatic activities. These have been extracted from the database of known structures, and a new protein structure can be scanned against the database to search for the presence of a known catalytic template. ASSAM and SPASM allow the user to define interesting residues in a protein structure and search similar sets of residues in the same relative spatial positions in other known structures. In a similar vein, Relibase is a database of protein–ligand complexes with sophisticated geometrical search facilities. All these tools are linked from the text WWW site, and form a formidable armory for functional analysis of new protein structures.

I5 STRUCTURAL ALIGNMENT

Key Notes

Structural alignment	It can be very difficult to find correct, biologically-meaningful alignments of very distantly related protein sequences because they contain only a very small proportion of identical monomers. In such cases, structural information can help because evolution tends to change structure less. Superimposing the backbones of similar structures implies structurally equivalent residues and this process is known as structural alignment.
Software	A variety of software is available for structural alignment, some of which is linked from the text WWW site.
Structural similarity	Structural alignment methods often produce measure of structural similarity. The most common of these is the RMSD, which is reported by most programs. This is the root mean square difference in position between the α carbon atoms of aligned residues in optimal structural superposition.
Structural similarity searches	Similarity searches of the structural database are available from several WWW sites.
Related topics	Sequence similarity searches (E1) Amino acid substitution matrices (E2) Database searches: FASTA and BLAST (E3)	Multiple sequence alignment and family relationships (F1) Classification of proteins of known three-dimensional structure: CATH and SCOP (I6)

Structural alignment

It should be obvious from the material in Sections E and F that the alignment of protein or nucleic acid sequences is a very important part of bioinformatics. An alignment defines a relationship between the elements of one sequence and those of another, and the degree of similarity of the sequences reflected in the alignment is the basic data on which we base judgments, by statistical or other means, of the likely biological meaning of the relationship. Proteins are likely to have similar structures and biological functions if their sequences can be aligned to show significant similarity.

Sequence alignment is relatively easy when the sequences are closely related, involving a high proportion of identical monomers, and few insertions and deletions. However, as we have commented earlier in this section, the sequences of evolutionarily and structurally related proteins can be very different. Divergent evolution can lead to sequences whose similarity is almost unrecognizable. In these cases, naïve sequence alignment based on information only from within the sequences and standard substitution matrices (Topics E1 and E2) can be very difficult. It can be almost impossible to find an alignment that is even approximately correct in biological or evolutionary terms.

Knowledge of protein structure provides a means by which highly divergent sequences can be aligned, and this process is known as **structural alignment**. This can be described conceptually as a process in which two similar structures are superimposed in three dimensions, so that peptide backbones of structurally equivalent residues lie close together in space. This superposition is then used to define a sequence alignment in which the aligned residues are structurally equivalent (see *Fig. 1*). Because structure is more strongly conserved during evolution than sequence, the structural alignment is much more likely to be correct in terms of biological function and evolution.

In *Fig. 2*, two alignments of a pair of distantly related globin sequences are shown – the first produced by sequence only means and the second a structural alignment. The alignments are different, and although the sequence only alignment shows a higher percentage of identical residues it is certainly not as good

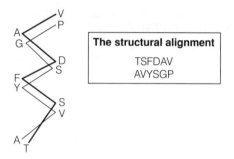

Fig. 1. *A conceptual view of structural alignment. Two protein three-dimensional backbone structures are shown optimally superimposed. Structure 1 has sequence TSFDAV and is represented by a thick line, and structure 2 has sequence AVYSGP and is represented by a thin line. The implied structural alignment is also shown with structurally equivalent residues aligned.*

(a)

```
SLSAAEADLAGKSWAPVFANKNANGLDFLVALFEKFPDSANFFADFK-GKSVADIKA-S
VLSPADKTNVKAAWGKVGAHAGEYGAEALERMFLSFPTTKTYFPHFDLSHGSAQVKGHG
                               *
PKLRDVSSRIFTRLNEFVNNAANAGKMSAMLSQFAKEHVGFGVGSAQFENVRSMFPGFVA
KKVADALTNAVAHVDDMPNALSALSDLHAHKLRVDPVNFKLLSHCLLVTLAAHLPAEFTP
                       *
```

(b)

```
XSLSAAEADLAGKSW-APVFANKN-ANGLDFLVALFEKFPDSANFF-ADFKGKSVA--DIK
V-LSPADKTNVKAAWGK-VGAHA-GEYGAEALERMFLSFPTTKTYFPHF-------DLS-H
                                 *
ASPKLRDVSSRIFTRLNEFVNNAANAGKMSA-MLSQ-FAKEHV-GFGVGSAQFENVRSM-F
GSAQVKGHGKKVADALTNAVAHV-D--DMPNAL--SALSDLHAHKLRVDPVNFKLLS-HCL
                                 *
PGFVASVAA--PPAGADAAWTKLFGL-IIDA-LKAAGA-
LVTLAAHLPAEFTPAVHASLDKFLASVST-VLT-SKY-R
```

Fig. 2. *Alignment of distantly related globin sequences (human hemoglobin, bottom; sea hare myoglobin, top) using (a) sequence only means (a local alignment of 87 residues showing 21% identical amino acids), and (b) structural alignment (an alignment of 139 amino acids showing 16% identical residues). The histidine residue that coordinates the heme iron is marked with an asterisk. Note these histidines are correctly aligned in (b) but not in (a).*

as the structural alignment. One very obvious advantage of the structural alignment is that the histidine residues that coordinate the heme iron in both globins are aligned (equivalenced), whereas they are not in the sequence only alignment.

Software

There is a large amount of software available to perform structurally based alignments, and some of this is linked from the text World Wide Web (WWW) site. A good example is DALI WWW (http://www.ebi.ac.uk/dali). The detailed methods used by the various computer programs are different, and almost all of the methods do not conform exactly to the simple conceptual description of structural alignment given above. Some methods make use of sequence as well as structure information, and almost all make use of information about secondary structure. There is no real consensus about which methods are the best, but any alignment of distantly related protein sequences will be better if created using structural information.

Structural similarity

Most structural alignment methods produce measures that can be used to assess the level of similarity of the aligned structures. A variety of such measures are used, many of which are specific to particular structural alignment methods. Nevertheless most methods report a measure known as **root mean square deviation (RMSD)**. When two structures have been aligned, they can be optimally superimposed so the aligned residues lie as close as possible to each other in three-dimensional space (*Fig. 1*). With the structures superimposed, distances can be measured between the α carbon atoms of the aligned residues, and the root mean square deviation is defined in terms of these distances by

$$RMSD = \sqrt{\frac{1}{N}\sum_i d_i^2}$$

where d_i is the distance between the ith pair of superimposed (aligned) α carbon atoms, and N is the number of such superimposed atoms (i.e. the number of aligned residues). This expression represents the square root of the mean square distance between aligned residues. If the structures were identical then the RMSD would be zero, because all residue α carbon atoms would lie directly on top of one another (zero distance apart), after optimal superposition. For less similar structures, the RMSD becomes larger. Generally RMSD values between 0.0 and 1.5 Å represent very similar structures, with higher values indicating progressively increasing structural dissimilarity. Clearly a small RMSD computed over a large number of residues (N) is more significant than a small RMSD computed over a small number of residues.

Structural similarity searches

Just as we are often interested in searching sequence databases for sequences similar to a query sequence, it is sometimes desirable to search the structural database for structures that are similar to a query structure. A number of WWW sites provide this type of search, including DALI, SSAP, TOPS, VAST and the Research Collaboratory for Structural Bioinformatics (RCSB) site (see the text WWW site for links). As with sequence searches, these search engines return a list of similar structures ranked according to some measure of similarity, such as RMSD.

I6 CLASSIFICATION OF PROTEINS OF KNOWN THREE-DIMENSIONAL STRUCTURE: CATH AND SCOP

Key Notes

Why classify protein structures?	Classification groups together proteins with similar structures and common evolutionary origins.
Example classifications	CATH (http://www.biochem.ucl.ac.uk/bsm/cath) and SCOP (http://scop.mrc-lmb.cam.ac.uk/scop)
Structural classes	Proteins can be assigned to broad structural classes based on secondary structure content and other criteria. CATH has four such broad classes, but SCOP uses more, giving a more detailed description of structural class.
Fold or topology	All classifications gather together proteins with the same overall fold or topology. Proteins in the same fold or topology class contain more or less the same SSEs, connected in the same way and in similar relative spatial positions.
Homologs and analogs	Homologs (homologous proteins) are related by divergent evolution from a common ancestor, and have the same fold. Analogs (analogous proteins) have the same fold, but other evidence for common ancestry is weak.
Super-folds	Super-folds are protein folds that seem likely to have arisen more than once in evolution. They are thought to have advantageous physicochemical properties. They appear in SCOP and CATH as fold or topology levels containing several homologous super-families.

Related topics	Multiple sequence alignment and family relationships (F1)	Conceptual models of protein structure (I1)
	Protein families and pattern databases (F2)	Obtaining, viewing and analyzing structural data (I4)
	Protein domain families (F3)	

Why classify protein structures?

To classify protein structures means to group them so that each group contains similar structures. In Topic F1, we learned that naturally occurring protein sequences can be grouped into evolutionary families, and in Topic I3, we discovered that protein structure is much more strongly conserved by evolution than protein sequence. To classify proteins by structural criteria is therefore the most powerful way to assigning them to families, and to reveal distant evolutionary relationships. Methods of protein structure classification rely heavily on the sequence comparison methods discussed in Section E and the structure comparison methods discussed in Topic I5.

Example classifications

CATH (http://www.biochem.ucl.ac.uk/bsm/cath) and **SCOP** (http://scop.mrc-lmb.cam.ac.uk/scop) are example classifications. Both are hierarchical (tree-structured), and their structures are shown in *Figs. 1* and *2*. Despite the apparent differences, the classifications agree to a large extent about which proteins should be grouped together.

Structural class

At the top level, each classification places protein structures into broad structural classes. CATH uses four classes reflecting secondary structure content: all α, all β, α and β (αβ), and few secondary structures. SCOP on the other hand has more divisions, separating out more small groups, like membrane and coiled-coil proteins, and also splitting the α and β class into two, α+β and α/β, depending on whether the α helices and beta strands are segregated in the fold (α+β) or mixed up and tending to alternate (α/β). The next level in CATH further separates the structural classes into broad architectures reflecting the overall shape of the protein, and has no equivalent in SCOP.

Fold or topology

The CATH T (topology) and SCOP fold levels separate proteins into groups that have the same overall fold. This means that the proteins have the same core **secondary structure elements (SSEs)**, connected and arranged in space in more or less the same way. The examples given in *Figs. 1* and *2* are part of the TIM barrel-fold level. Within this level, all proteins have the well-known TIM barrel structure: a parallel eight-stranded beta barrel surrounded by α helices. When proteins have the same fold or topology, they can usually be structurally aligned to give large sections of superimposed backbone structure with low root mean square deviation (RMSD; Topic I5) values.

Homologs and analogs

It is thought that many proteins with the same fold have emerged by divergent evolution from a common ancestor, but it is equally possible that they have no common ancestor and adopt the same fold simply because that fold is favorable from a physicochemical point of view. In the former case, the proteins are called

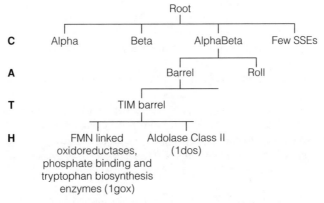

Fig. 1. The structure of the CATH classification illustrating the positions of the glycolate oxidase structure with PDB code 1gox (Topic C4) and the class II aldolase structure 1dos, which share the TIM barrel fold. 1gox and 1dos share the AlphaBeta broad structural class (C), barrel architecture (A) and TIM barrel topology or fold (T). They belong to different homologous super-families (H).

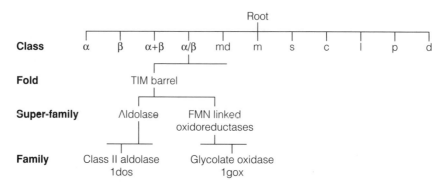

Fig. 2. The structure of the SCOP classification illustrating the positions of the FMN linked oxidoreductase and aldolase super-families, and the classification of 1gox and 1dos to compare with Fig. 1. At the class level, SCOP has more divisions than CATH: α, all α; β, all β; α+β, α and β segregated; α/β, α and β alternating; md, multi-domain; m, membrane and cell surface; s, small proteins; c, coiled coil proteins; l, low resolution structures; p, peptides; d, designed proteins.

homologs and are said to be **homologous**, in the latter they are called **analogs**, and are said to be **analogous**. It is almost impossible to know whether two weakly similar proteins have derived from a common ancestor or not. However, CATH and SCOP divide proteins with the same fold into homologous super-families. Within such a super-family, there is good evidence that the proteins are homologous. This evidence might be striking similarity in sequence, structure or function. When proteins with the same fold are in different super-families, it means that evidence for homology beyond the shared fold is weak, or nonexistent.

Figures 1 and *2* show that although the two classifications agree on the fold classification of the proteins in question (TIM barrel), they differ slightly at the homologous super-family level. Both assign the glycolate oxidase and aldolase examples to different homologous super-families, but CATH has a separate super-family for the class II aldolases, which are gathered into a general aldolase super-family in SCOP (containing class I aldolases in addition). This reflects a difference of expert opinion on whether there is sufficient evidence that the two classes of aldolase should be viewed as homologous. It should also be noted that the authors of these classifications do occasionally change their minds about the classification of particular groups of proteins at this level, and this is often reflected in changes made in newer versions of the database.

Super-folds

Only a relatively small number of the different folds in CATH or SCOP contain more than two homologous families. Such folds appear to have arisen more than once in evolution, and are called **super-folds**. Example super-folds are the TIM barrel and the immunoglobulin fold. Characteristics of the super-folds are that they tend to exhibit approximate symmetries, and are characterized by repeated super-secondary structures. These suggest the possible physicochemical favorability of these folds.

Families

Proteins within homologous super-families are often quite distantly related. Both classifications divide each super-family into a number of smaller families, within which the relationships between the proteins is much stronger, and often apparent at the level of sequence similarity.

I7 INTRODUCTION TO PROTEIN STRUCTURE PREDICTION

Key Notes

Why predict structure?

Structure prediction is interesting because experimental structure determination is still much slower than sequence determination. Structure predictions help us to understand function and mechanism and can be useful for rational drug design. The early work of Levinthal and Anfinsen made structure prediction a fascinating scientific problem.

What is structure prediction?

We will discuss methods in the categories of comparative modeling, fold recognition, secondary structure prediction, *ab initio* prediction and transmembrane segment prediction.

Related topics

Conceptual models of protein structure (I1)
Structure prediction by comparative modeling (I8)

Secondary structure prediction (I9)
Advanced protein structure prediction and prediction strategies (I10)

Why predict structure?

In the previous topics in this section, we have discussed protein three-dimensional structures and how these enable proteins to carry out their functions. An understanding of structure leads to an understanding of function and mechanism of action. Structural knowledge is therefore vital if we are to proceed to a complete understanding of life at the molecular level. Unfortunately, structural knowledge is still rather limited, because the experimental process of structure determination is slow for most proteins and not possible for many. The database of known structures currently contains more than 18 000 protein structures, but the databases of sequences contain hundreds of thousands of sequences. This gap between sequence and structure knowledge is often termed the **sequence structure gap**; from a practical point of view, it is the main factor motivating the need for predictions of protein structure. To be added to this is the fact that many pharmaceutical drugs act selective binding to target proteins, and knowledge of protein structures can aid the process of rational structure-based drug design (the design of drug molecules based on the structures of the proteins with which they are intended to interact; see Section N).

As well as having enormous practical significance, structure prediction is a fascinating scientific problem that has interested scientists since before the first protein structures were determined. In fact, Pauling predicted the structure of the α-helix before it was observed experimentally. Much of the fascination of the problem stems from the work of two scientists, Anfinsen and Levinthal. By a set of elegant experiments on RNase A, Anfinsen was able to prove that proteins can fold to their native structures spontaneously, without the intervention of any other agent. The protein fold is therefore somehow coded in the amino acid sequence. Levinthal, on the other hand, pointed out that even relatively small proteins have an astronomically large number of possible structures, and that the process of finding the correct one cannot possibly proceed by a random search of the possibilities because this would simply take too long. To discover how the protein sequence codes for a three-dimensional structure, and to understand how this structure folds has therefore been a problem of much scientific interest for many years. Despite this, the problem is still far from solved, and even though there has been progress, we still do not know in general terms how structure is encoded in sequence.

What is structure prediction?

In the most general form, structure prediction means to make a prediction of the relative position of every protein atom in three-dimensional space using only information from the protein sequence. However, not all prediction methods are so general, and some predict limited aspects of structure, for instance secondary structure, without proceeding to predictions of actual atomic positions. In the Topics that follow, we will discuss protein structure prediction methods in four categories, the use of comparative modeling, fold recognition, secondary structure prediction and *ab initio* prediction. We will also discuss the prediction of membrane spanning segments and integral membrane protein topology.

It is useful to categorize prediction methods further according to their theoretical basis, either **ab initio** or **knowledge-based**. Like all physicochemical systems, proteins are believed fold to attain a state of minimum thermodynamic free energy. In our terminology, only methods that attempt to calculate and minimize this free energy, or a suitable approximation, are *ab initio* methods. They proceed from fundamental physical principles, using the accepted theories of quantum mechanics and statistical thermodynamics, to predict protein structures. This is very difficult. Proteins, when modeled with enough solvent molecules to be realistic, are enormous systems with many thousands of atoms. With current technology, detailed calculations of exact free energies are just not possible, and are not likely to be for many years to come. The task therefore is to find a suitable approximation to the free energy that captures the essentials of the folding problem. Such an approximation has not yet been found.

In contrast to *ab initio* methods, knowledge-based methods attempt to predict protein structure using information taken from the database of known structures. The simplest example of this is to predict that a sequence, that is similar to a sequence of known structure, will adopt that same basic structure. This was shown by Sander and Schneider (Topic I3), but there are many other ways of using information from the database of known structures in predictions. The methods of comparative modeling (Topic I8) and fold recognition (Topic I10) are knowledge-based methods. We also view secondary structure prediction (Topic I9) as a knowledge-based method. This is because most secondary structure prediction methods have been trained on data from known structures. However, it is sometimes considered as an *ab initio* method because it is applicable even to sequences that form tertiary structures that have not yet been observed in any structural relative, however distant. There is no doubt that knowledge-based methods are currently the most accurate and practically useful protein structure prediction methods.

In the past few years, protein structure prediction methods have been subjected to some rigorous blind testing associated with the **CASP (Critical Assessment of Structure Prediction)** competitions. The results of these competitions have added enormously to our understanding of the accuracy of the various methods. Details of the CASP competitions are linked from the text WWW site at http://predictioncenter.llnl.gov/.

I8 STRUCTURE PREDICTION BY COMPARATIVE MODELING

Key Notes

Theoretical basis	Sequences with more than 25% identity over an alignment of 80 residues or more adopt the same basic structure. This is the basis of prediction by comparative modeling.
Ingredients	All that is needed is an alignment between a sequence of unknown structure (target) and one or more of known structure (template(s)) with the above property. Template structures can be found by standard sequence similarity search methods. Lack of suitable template structures is the main limitation of the method, but structural genomics projects are likely to change this in coming years. The accuracy of the alignment is crucial if a good prediction is to be obtained.
The process	Known structure(s) (templates) are used as the basis of the prediction. The process can then be viewed conceptually as comprising placement of conserved core residues, modeling of variable loops, side-chain positioning and optimization, and model refinement. Conserved residues and some side-chain positions can be obtained directly from structural information in the templates. Modeling of variable loops often makes use of the spare parts algorithm, and there are sophisticated algorithms for side-chain placement to obtain an optimally packed hydrophobic core.
Accuracy	Accuracy is controlled almost entirely by the quality of the alignment. Good alignments yield good predictions with most of the main software packages. Of all prediction methods, comparative modeling produces the most accurate models.
Availability	Comparative modeling is available on the WWW and in several free and commercial software packages. Some of these are linked from the text WWW site.

Related topics	Structural alignment (I5)	Secondary structure prediction (I9)
	Introduction to protein structure prediction (I7)	Advanced protein structure prediction and prediction strategies (I10)

Theoretical basis　　In Topic I3, we discovered that two proteins share the same structure if their sequences can be aligned to show at least 25% identity in an alignment of 80 or more residues. This is the basis of the method of **comparative modeling** (also sometimes known as **homology modeling**). If a sequence of unknown structure (usually called the **target sequence**) can be aligned with one or more sequences of known structure, so as to satisfy this condition, then the known structures can be used to predict the structure adopted by the target sequence. The proteins of known structure used are usually referred to as **template structures**.

Ingredients

All that is necessary for comparative modeling is an alignment between the target sequence and the sequences of the template structures. The first step is to locate possible template structures, and this can be done using the standard sequence similarity search methods discussed in Topic E3. In this case it is only necessary to search the database of sequences whose structures are already known by experimental means. This is a relatively small database, and the lack of availability of suitable template structures is the main limitation to the use of comparative modeling at the present time. This situation may change in the coming years. Large-scale projects aimed to create high throughput structure determination, under the name of **structural genomics**, are now underway (see for example the New York Structural Genomics Consortium, http://www.nysgrc.org/). It is likely that suitable template structures will become available for most protein sequences within the next 10–15 years.

When template structures have been obtained it is necessary to align their sequences with the target sequence using a multiple alignment tool (Topic F1). From a user's point of view, the alignment process is the most crucial step in comparative modeling. If the alignment is good, then a good model will result; if it is poor, then the model will also be poor. When the target and template sequences are closely related, with high percentage identities (for example 70% and above) then automated alignment methods usually produce good alignments. Nevertheless, even when percentage identity is very high, the alignment should be inspected for any obvious alignment errors. For example, it is advantageous to check that conserved key structural and functional residues are aligned. If there are key structural cysteines, glycines or prolines (Topic I3), then these should be aligned with residues of corresponding types in all the sequences.

When target and template sequences are more weakly related, with lower percentage identities, then alignment errors are more common and more manual checking is necessary. It may be very difficult to identify the correct alignment, and this can result in poor predictions. There are now many software packages that do completely automated comparative modeling, with no manual user input. These function well when target and templates are closely related, but should be treated with caution when relationships are more distant. Always check that the alignment used was acceptable from a structural and functional perspective.

The process

The various software packages carry out the comparative modeling process in slightly different ways, and what we describe here are generic features of the process that are common to all methods. A schematic illustration of the process is given in *Fig. 1*. The process begins by analysis of the template structures. If there is more than one such structure, an average structure is often calculated by superposing the structures in three dimensions (Topic I5) and calculating average atomic positions. The contributions of the various templates in this process are often weighted according to their degree of similarity to the target sequence, with more similar sequences more strongly weighted. In this process a framework of template atomic positions is calculated.

Next, using the alignment of the target sequence to the templates, the structure is divided into two distinct types of region, the structural core and the nonconserved loops. The core can be considered to comprise mainly the

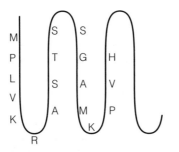

Sequence of known structure:

MPLVKRASTSSGAMKPVH...

Schematic of known 3D structure (template)

Sequence of unknown structure (target):

MPILKRGTSTSYGAMRPIY...

Aligned with sequence of known structure:

```
MPILKRGTSTSYGAMRPIY
MPLVKRASTSS_GAMKPVH
```

Predict the 3D structure of the target sequence by replacing the old residues on the known structure with those from the target sequence made equivalent by the alignment.

Need to position new side chain atoms.

Small structural changes where gaps are found in the alignment (loop modelling, dotted circle).

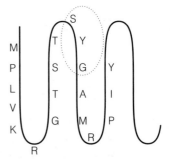

Predicted 3D structure

Fig. 1. Comparative modeling.

secondary structure elements (helices and strands) of the templates, within which each residue is aligned with a residue from the target sequence. The nonconserved loops are mainly surface loops for which, according to the alignment, target and template sequences have different numbers of residues (i.e. there are gaps in the alignment). Structure prediction for the core is easy, because the backbone atomic positions from the averaged template structure can be used directly as backbone atomic positions for the predicted (model) structure of the target sequence.

Within nonconserved loops, structure prediction is more difficult, and more sophisticated structure prediction methods have to be used. The simplest method of loop structure prediction is to use the **spare parts** algorithm. This method makes use of a database of known loop structures from other proteins that are not necessarily similar in sequence to the target. For each loop needed, a spare part loop is found in this database that fits into the gap in the modeled structure, and this is taken as the predicted structure for the loop region. There are more sophisticated loop prediction algorithms, involving more complex calculation and minimization of estimated loop energies, but the simple spare parts algorithm seems to work well in practice. Prediction accuracy in the region of predicted loops is generally much lower than it is in the core, and while most loop prediction algorithms perform well on short loops (less than six residues) they become increasingly inaccurate as loop length increases.

When the steps above are complete there are predicted positions for all the backbone atoms of the target sequence, and the process turns to the question of **side-chain positions**. Some side-chain identities are conserved between target and template structures, and the structural information from the templates can be used to place these side-chains in the predicted structure. However, not all side-chains are conserved and the positions of the atoms in these need to be predicted. This is usually done by choosing from a library of allowed side-chain structures (sometimes called a **side-chain rotamer library**), so that the atoms of the side-chains fill the available space in the protein interior without being too close (clashing) with any other protein atoms. The algorithms used to do this are often quite sophisticated, and aim to produce a protein hydrophobic core that is tightly packed, like those observed in experimentally determined structures.

At this stage, the preliminary structural model for the target sequence is complete and many predictions stop at this point. Sometimes, however, it is useful to refine the model further. Typically this involves the use of energy minimization software that can make small changes to atomic positions in order to produce a slightly lower energy model. This can be advantageous, for instance to move any atoms that were left too close in proximity after the modeling process above, but the advantage in terms of the accuracy of the model is debatable.

Accuracy

We have already commented that the accuracy of comparative models is determined almost entirely by the accuracy of the alignment. When alignments are good, which they typically are when the sequences are very closely related, then very accurate models are possible. Accuracy is typically measured in terms of root mean square deviation (RMSD; see Topic I5 for a definition) between the α carbon positions in the predicted structure and the actual structure of the target sequence. RMSDs of less than 1.0 Å represent very good predictions; this degree of difference between the two structures is similar to the degrees of difference between two separate experimental determinations of the same protein structure. When the percentage sequence identity between template structures and target sequence exceeds 70% it is reasonable to expect that the model should be accurate to an RMSD of less than 2–3 Å, even using completely automated methods. When the percentage identity drops below 40% then getting a good alignment, often with manual intervention, becomes crucial and automated methods can fail very badly.

It is of course not possible to know how accurate a model is without an independent experimental determination of the actual structure, but the above guidelines should be borne in mind. It is also possible to check some aspects of the predicted structure; for instance, software described in Topic I4 for checking of stereochemical parameters of the structure will identify any unusual bond lengths or angles that might indicate potential problems in the prediction. Predicted structures should be viewed in general as quite low-resolution models. For applications where a very accurate structure is required, for example detailed structure-based drug design, they are probably too low in accuracy, particularly in respect of nonconserved loops and side-chain positions. Nevertheless, if it is possible to construct a structure prediction by comparative modeling then this prediction is likely to be more accurate than a prediction by any other method.

Availability

Comparative modeling is available on the World Wide Web (WWW) at the SWISS-MODEL site, and using the SWISS-PDBVIEWER software (http://www.expasy.ch/swissmod/SWISS-MODEL.html). It is also available in several commercial software packages.

I9 SECONDARY STRUCTURE PREDICTION

Key Notes

What is secondary structure prediction?	Secondary structure prediction predicts the conformational state of each residue in three categories, helical, extended or strand, and coil. Many methods are based on ideas related to secondary structure propensity, which is a number reflecting the preference of a residue for a particular secondary structure. Early methods had accuracies of around 60% (the percentage of residues predicted in the correct helical/extended/coil state). Examples of early methods are the Chou–Fasman rule-based method and the information-theoretical GOR method.
Multiple sequence information	Using multiple alignments of related sequences can improve prediction accuracy enormously by revealing patterns of conservation indicative of certain secondary structures.
Accuracy of state-of-the-art methods	Currently methods claim an average accuracy over trusted test sets of proteins equal to more than 70% of residues correctly predicted. This increase in accuracy can be attributed to the availability of more structural data, and the use of more sophisticated algorithms or methods.
Prediction of transmembrane segments	Membrane-spanning segments in integral membrane proteins can be predicted with reasonable accuracy. Most methods make use of a search for contiguous runs of hydrophobic residues that span a lipid membrane. Some methods also predict the orientation (in–out) or topology of the membrane-spanning segments, but this is usually less accurate.
Availability of tools	Most of the secondary structure and transmembrane segment prediction tools mentioned in this topic are available from the ExPASy WWW site (http://www.expasy.ch).

Related topics	Determination of protein structure (B2)	Structure prediction by comparative modeling (I8)
	Conceptual models of protein structure (I1)	Advanced protein structure prediction (I10)
	Introduction to protein structure prediction (I7)	

What is secondary structure prediction?

The previous topic introduced comparative modeling as the method of choice for protein structure prediction. When suitably related template structures do not exist for a particular target sequence, secondary structure prediction is a viable alternative. Unlike comparative modeling, it does not produce a full atom model of the tertiary structure, but rather just provides a prediction of the secondary structure state of each residue, either **helical**,

strand or **extended**, or **coil**. The predictions are sometimes known as **three-state predictions**.

Most methods of secondary structure prediction have been trained on the database of known structures. Two early methods were those of Chou and Fasman, and GOR (Garnier–Osguthorpe–Robson, after its inventors). Although these methods work in different ways, the former being rule based and the latter using information-theoretical ideas, they both make use of the idea of **secondary structure propensity**. The amino acids seem to have preferences for certain secondary structure states, which are shown in *Table 1*. For instance, glutamic acid has a strong preference for the helical secondary structure, and valine a strong preference to be in strands. Glycine and proline have lower than average propensity for both types of regular secondary structure, reflecting a tendency to be found in loops, and the some hydrophobic amino acids (e.g. phenylalanine) have strong preferences for both types of secondary structures reflecting their tendency to make up the structural core. However, none of the preferences are particularly strong, and all amino acids are found in each type of secondary structure quite often. This means that predictions of secondary structure cannot be based on single residues.

In the Chou–Fasman method, a helix is predicted if, in a run of six residues, four are helix favoring, and the average value of the helix propensity is greater than 1.0 and greater than the average strand propensity. Such a helix is extended along the sequence until a proline is encountered (these are known to break helices) or a run of four residues with helical propensity less than 1.0 is found. A strand is predicted if, in a run of five residues, three are strand

Table 1. Helical and strand propensities of the amino acids. A value of 1.0 indicates that the preference of that amino acid for the particular secondary structure is equal to that of the average amino acid, values greater than one indicate a higher propensity than the average, and values less than one a lower propensity than the average. The values are calculated by dividing the frequency with which the particular residue is observed in the relevant secondary structure by the frequency for all residues in that secondary structure

Amino acid	Helical (α) propensity	Strand (β) propensity
GLU	1.59	0.52
ALA	1.41	0.72
LEU	1.34	1.22
MET	1.30	1.14
GLN	1.27	0.98
LYS	1.23	0.69
ARG	1.21	0.84
HIS	1.05	0.80
VAL	0.90	1.87
ILE	1.09	1.67
TYR	0.74	1.45
CYS	0.66	1.40
TRP	1.02	1.35
PHE	1.16	1.33
THR	0.76	1.17
GLY	0.43	0.58
ASN	0.76	0.48
PRO	0.34	0.31
SER	0.57	0.96
ASP	0.99	0.39

favoring, and the average value of the strand propensity is greater than 1.04 and greater than the average helix propensity. Such a strand is extended along the sequence until a run of four residues with strand propensity less than 1.0 is found. This is a simple rule-based method dependent on finding runs of residues with preference for one type of secondary structure. The GOR method is different, considering the information carried by a residue about its own secondary structure, in combination with the information carried by other residues in a local window of eight residues on either side in the sequence of the residue concerned.

The accuracy of these early methods based on the local amino acid composition of single sequences was fairly low, with often less than 60% of residues being predicted in the correct secondary structure state. This should be viewed in light of the fact that for proteins containing roughly equal proportions of helix, strand and coil then the accuracy of random predictions would be 33% (i.e. the correct state would be predicted for one in every three residues).

Multiple sequence information

In the late 1980s, it was realized that the evolutionary information contained in multiple sequence alignments could significantly improve secondary structure predictions. Several factors contribute to this increase in accuracy, but they all relate to the patterns of conservation revealed in the multiple alignment. First, a residue with a high propensity for a particular secondary structure in one sequence may have occurred by chance, but if it is part of a conserved column in which all the residues have high propensity for that type of secondary structure then this provides much more predictive evidence. Second, multiple alignments can reveal more subtle patterns of conservation. For instance, a large proportion of α-helices in globular proteins are amphipathic. These helices (see Topic I3) have hydrophobic and hydrophilic faces and are associated with periodic patterns of hydrophobic and hydrophilic residues in the sequence. The appearance of conserved patterns of this type is therefore highly predictive of α helical structure. Finally, as we commented in Topic I3, insertions and deletions are much more likely in surface loops (coil), so regions of the alignment where these are common are more likely to be in the coil state. The use of evolutionary information from multiple alignments improved the accuracy of secondary structure prediction methods significantly, resulting in methods capable of an accuracy of around 66% at the time.

Accuracy of state-of-the-art methods

Since the late 1980s, we have seen an explosion in the amount of data in the sequence and structure databases that has increased the amount of evolutionary and structural information available to secondary structure prediction algorithms. This, coupled with improvements to the algorithms themselves, has led to much increased accuracy. Examples of state-of-the-art prediction methods are PHD (Rost and Sander), DSC (King and Sternberg), PREDATOR (Frishman and Argos), NNSSP (Salamov and Solevyev) and PSI-PRED (Jones). These methods use a variety of techniques to make their predictions, many of which are based in the fields of artificial intelligence or machine learning. PHD and PSI-PRED make use of artificial neural networks (simple computational models of the neural networks in the brain), and DSC is based on a machine learning method known as linear discriminant analysis. NNSSP (Nearest Neighbor Secondary Structure Prediction) makes predictions based on locating the most similar sequence segments in the database of known structures. The accuracy of all these methods is above 70%, and, depending on the test set of predictions chosen, accuracies of above 75% have been achieved. An example secondary

```
Conf: 9688223888978378999999999999986578877899999861597414786664032
Pred: CEEEECCCCCCCCHHHHHHHHHHHHHHHHHCCCCCCHHHHHHHHHCCCHHHCCHHHHHHCC
Act : NNNNNNNNNNNNNNCHHHHHHHHHHHHHHHHHCCCCCCHHHHHHHHHCCHHHHCCHHHHHHHC
 AA: MHLYSSDFPLMMDEKELYEKWMRTVEMLKAEGIIRSKEVERAFLKYPRYLSVEDKYKKYA
              10        20        30        40        50        60
```

```
Conf: 237776523785312767999999972689889899976370689999996319879998
Pred: CCCCCCCCCCCCEECCHHHHHHHHHHHHCCCCCCEEEEEECCCHHHHHHHHHHHCCCCEEEEE
Act : CCCCCEECCCCCEECCHHHHHHHHHHHHCCCCCCCEEEECCCCCHHHHHHHHHHHCCCEEEE
 AA: HIDEPLPIPAGQTVSAPHMVAIMLEIANLKPGMNILEVGTGSGWNAALISEIVKTDVYTI
              70        80        90       100       110       120
```

```
Conf: 229999999999988659987426765443456657896889987972311367898706
Pred: ECCHHHHHHHHHHHHHHCCCCCEEEECCHHHCCCCCCCCCCEEEECCHHHCCHHHHHHCC
Act : ECCHHHHHHHHHHHHHHCCCCCEEEEECCHHHCCHHHCCEEEEEECCCCCCCCHHHHHCEE
 AA: ERIPELVEFAKRNLERAGVKNVHVILGDGSKGFPPKAPYDVIIVTAGAPKIPEPLIEQLK
             130       140       150       160       170       180
```

```
Conf: 799999845797860689999626985568852555666152433541332522 9
Pred: CCCEEEEECCCCCCEEEEEEEEECCEEEEEEEECCEEEEECCCCCCCCHHHHHCC
Act : EEEEEEEEECCCCCCEEEEEEEEECCEEEEEEEEEECCCCCCCCCCCCNNNNNNNN
 AA: IGGKLIIPVGSYHLWQELLEVRKTKDGIKIKNHGGVAFVPLIGEYGWKEHHHHHH
             190       200       210       220       230
```

Fig. 1. Secondary structure prediction. The PSI-PRED server was used to predict the secondary structure of a protein (L-isoaspartate O-methyltransferase) whose actual three-dimensional structure had very recently been determined. Shown above the sequence (AA) are the actual secondary structure (Act), the predicted secondary structure (Pred) and a measure of the confidence of prediction [0 (uncertain)–9 (very confident)]. The percentage of residues predicted in the correct secondary structure is 83%. C, random coil; H, helix (α or 3_{10}); E, extended (β); N, structure unknown.

structure prediction from PSI-PRED in shown in *Fig. 1*. This prediction was carried out using a protein of very recently determined structure, and the actual secondary structure from that experiment is shown for comparison with the predicted structure. The overall accuracy in this case is higher than average (83%). It should be noted that many of the prediction errors are residues on the ends of helices or strands and that the secondary structure of these residues may not be well defined in the protein: it can be very difficult to decide whether a residue at the end of a helix is part of the helix or not.

One caveat that must always be born in mind is that all secondary structure prediction methods have been trained using structures from the database of known protein structures. This set of proteins is biased for various reasons; some very common proteins (for instance integral membrane proteins) are under-represented. The prediction methods can be expected to work better for proteins that are in some sense similar to the proteins in this database, and less well for others. It makes no sense therefore to apply standard secondary structure prediction methods to integral membrane proteins.

Prediction of transmembrane segments

In Topic I1 we described the structure of integral membrane proteins as consisting of segments (usually helical) spanning the lipid membrane, connected by loops lying outside the membrane. The membrane-spanning segments tend to contain a high proportion of hydrophobic residues and are often more than 20 residues in length, corresponding to six to seven helical turns for transmembrane helices. Such relatively long runs of predominantly hydrophobic residues are seldom found in water-soluble globular proteins. This means that it is often possible to predict, based on the runs of hydrophobic residues, whether or not a protein is an integral membrane protein, and if so, where the membrane-spanning segments are in the sequence.

Tools have been developed to predict transmembrane segments, including TMPred, TMHMM and TopPred. Many of these tools also predict the membrane-spanning topology. This is a prediction of the orientation of the helices with respect to the membrane. For instance and i→o (in to out) helices have their N-termini inside whatever cell or organelle the membrane bounds, and their C-termini outside. Clearly, a consistent prediction of topology must have i→o helices followed by o→i helices and *vice versa*. Predictions of the membrane-spanning topology are generally less reliable than predictions membrane-spanning segments irrespective of orientation.

An example of a prediction of membrane-spanning segments is shown in *Table 2*, where predictions have been made for the much-studied G protein-coupled receptor rhodopsin. In this case the predictions are quite accurate, with all seven transmembrane segments predicted in the correct sequence. This is of course not always the case, and it is quite common for some real membrane-spanning segments not to be predicted (false negatives), or for some non-membrane-spanning segments to be predicted as membrane spanning (false positives). False negatives often occur in ion channel proteins, where some of the membrane-spanning helices have hydrophilic sides, which cluster together to make a hydrophilic path through the membrane along which charged ions can pass. The extra hydrophilic residues in these helices often mean that they evade detection by methods searching for hydrophobic residues. False positives are often found in secreted proteins that contain an N-terminal signal peptide. Such signal peptides tend to be hydrophobic, and this sometimes leads to their confusion with membrane-spanning segments.

Table 2. Prediction of transmembrane helical segments for human rhodopsin (a G protein-coupled receptor). Predicted and actual positions refer to residue numbers with respect to the amino acid sequence in SWISS-PROT, where the actual positions of helices are recorded. Predictions were made with TopPred

Helix	Predicted position	Actual position	Predicted orientation	Actual orientation
1	37–63	37–61	o→i	o→i
2	74–99	74–98	i→o	i→o
3	115–140	114–133	o→i	o→i
4	153–175	153–176	i→o	i→o
5	203–221	203–230	o→i	o→i
6	253–276	253–276	i→o	i→o
7	286–309	285–309	o→i	o→i

Availability of tools

Most of the secondary structure and transmembrane segment prediction tools mentioned in this topic are available from the ExPASy WWW site (http://www.expasy.ch), under proteomics tools.

I10 ADVANCED PROTEIN STRUCTURE PREDICTION AND PREDICTION STRATEGIES

Key Notes

Fold recognition

Fold recognition aims to detect very distant structural and evolutionary relationships. It aims to detect when a protein adopts a known fold even if it does not have significant sequence similarity to any protein of known structure. Methods generally try to find the most compatible fold in a library of known folds using both sequence and structural information. An alternative term for fold recognition is threading.

Ab initio prediction

These methods rely on first principles calculation and are not yet sufficiently well developed to be of real use in practical structure prediction.

Strategies

After a thorough preliminary sequence analysis using the methods of Sections E and F, the best strategy employs first comparative modeling, and if not successful, secondary structure prediction followed by fold recognition.

Related topics

Sequence similarity searches (E1)
Iterative database searches and
  PSI-BLAST (E5)
Multiple sequence alignment
  and family relationships (F1)

Introduction to protein structure
  prediction (I7)
Structure prediction by comparative
  modeling (I8)
Secondary structure prediction (I9)

Fold recognition

In Topics I3 and I6, we discussed the evolution of protein structures and established that similarity in sequence is sufficient to guarantee similarity in structure. The method of comparative modeling is based on this fact. However, we also established that during evolution, struture is much more strongly conserved than sequence. There are many cases of distantly related proteins with the same structure but whose level of sequence similarity is well below the 25% needed to guarantee structural similarity and the possibility of comparative modeling. Fold-recognition methods are about detecting these distant relationships, and separating them from chance sequence similarities not associated with a shared fold. They operate by searching through a library of known protein structures (called a **fold library**) and finding the one most compatible with query sequence whose structure is to be predicted. An alternative name for fold recognition is **threading**.

The product of fold recognition is usually an alignment between the query sequence and one or more distantly related sequences of known structure. Fold recognition can therefore be viewed as an extension of the comparative modeling method to very distant relationships. Once the alignment between the sequence and the distantly related known structures has been obtained, a full three-dimensional structure of the protein to be predicted can be obtained using the usual methods of comparative modeling. Because relatively few protein structures are known, most new sequences cannot have their structures predicted by ordinary comparative modeling. Fold recognition is valuable because it has the potential to extend significantly the number of protein sequences whose structures can be predicted.

In Topic I6 we made they distinction between homologous proteins which share the same fold because they have evolved by divergent evolution from a common ancestor, and analogous proteins which share the same fold perhaps for physicochemical reasons with little evidence for common ancestry. Fold-recognition methods were originally intended to detect both analogs and homologs of the query sequence, but current evidence suggests that they can detect distant homology but not analogy.

Fold-recognition methods employ a variety of approaches, most often mixing the similarity in sequence detected by amino acid substitution matrices with further structural information. For instance the **3D-PSSM (three-dimensional Position Specific Scoring Matrix)** uses fold library structures described in terms of ordinary one-dimensional sequence profiles generated by PSI-BLAST (Topic E5), and also 3D profiles. The information within the 3D profile includes secondary structure and solvation potential. The solvation potential takes account of the tendency of hydrophobic amino acids to occupy structural positions that are not accessible to solvent (buried in the hydrophobic core). The secondary structure component measures the degree of similarity between the predicted secondary structure of the query sequence, using one of the methods described in Topic I9, and the secondary structure of the fold library member. There is very good evidence that the inclusion of structural information like secondary structure and solvation results in methods that are better able to detect distant homology than the most sensitive sequence only methods (for instance PSI-BLAST), and therefore that these are useful in structure prediction. Several fold-recognition methods operating in related ways to 3D-PSSM are linked from the text WWW site. The URL for 3D-PSSM is http://www.sbg.bio.ic.ac.uk/~3dpssm/.

One problem with fold-recognition methods is that while they are often able to recognize a distant homology to a known fold, they sometimes do not provide particularly accurate alignments to that fold. This situation is improving as the methods develop, but it must always be borne in mind that alignments may not be very accurate. Inaccurate alignments lead to poor structural predictions, using the comparative modeling methods of Topic I8.

Ab initio prediction

Ab initio methods were introduced in Topic I7 as methods that attempt to predict protein structures from first principles using theories from the physical sciences like statistical thermodynamics and quantum mechanics. We will say little more about them here, except to set them in context of the methods we have already described. There is no doubt that first principles predictions are very attractive from an intellectual viewpoint. However, these methods are still not sufficiently well developed to be of real use in practical protein structure prediction. Proteins in their natural solvation environments are very large systems, and are currently beyond the scope of accurate calculation with accepted theories. This is not to say that research in *ab initio* methods is uninteresting, indeed to understand the process of protein folding in this way would be a huge scientific achievement; but, from the point of view of the practicing molecular biologist or biochemist, the methods are currently of little use by comparison with the methods of comparative modeling, secondary structure prediction and fold recognition described previously.

Strategies

We close this section by outlining a strategy for protein structure prediction for a new query sequence of unknown structure. The first step should be perhaps to identify any features of the sequence that might affect the strategy to be adopted. The sequence should be examined for potential membrane spanning segments that might indicate the presence of an integral membrane domain. There are now some integral membrane proteins of known structure, but not many, and structure prediction for such domains remains difficult. Other sequence features that should be identified could be regions of low compositional complexity (Topic E4), since few of these appear in the database of known structures and therefore structure prediction might be inaccurate in such regions. The presence of coiled coils could be tested (Topic E4). Finally an analysis of the sequence by a tool like Interpro (Topic F3) might reveal the presence of known domains and perhaps the overall domain structure of the sequence. This could be supplemented by a PSI-BLAST search (Topic E5), which might reveal other related sequences and the sub-sequences (domains) where they are related. If the protein is multi-domain, it would make sense to predict each domain separately, if the positions of domains in the sequence could be defined.

The most accurate and comprehensive structure prediction method is comparative modeling, so the first prediction step of the strategy is to see if this is possible. This requires the identification of sequences of known structure with sufficient similarity to the query sequence, and can be carried out at the SWISS-MODEL WWW site. If this search is successful, then a comparative model can be built and there is no need for any further action. All other structure prediction methods are less accurate and less useful than comparative modeling. Even to make a secondary structure prediction is not useful, because the secondary structure of the comparative model itself is likely to be more accurate than any other secondary structure prediction method.

Comparative modeling is only possible for a minority of new protein sequences. When it is not possible, the next logical step is secondary structure prediction. This can be applied to any sequence, but of course is more accurate for globular protein domains and likely to be less accurate for integral membrane domains and low complexity regions. This will give a prediction of helix, strand or coil for each residue, which might be of use, for example in mutagenesis studies. Following secondary structure prediction, fold-recognition methods might give an idea of how the secondary structures pack into the tertiary fold, but these methods should be applied with some caution. It must be remembered that fold recognition detects very distant relationships where even structural divergence can be significant, and that 3D structure predictions from this method are generally much less accurate than those from standard comparative modeling.

J1 MICROARRAY DATA: ANALYSIS METHODS

Key Notes

Raw data from microarrays	Microarray data comprise images from hybridized arrays representing hybridization signal intensities for individual spots. These may be generated by single fluorescent, dual fluorescent, radioactive or colorimetric labels and the recording methods differ in each case.
Data quality	It is essential to record signal intensities from individual spots accurately as errors in data recording cannot be detected or corrected at a later stage. Software for reading microarrays is generally provided with the recording equipment (scanner or phosphorimager) but manual adjustment is necessary to compensate for variations in array manufacture. The signal must be corrected for background (nonspecific hybridization, autofluorescence, contamination) and hybridization controls must be used when comparing results across different arrays.
Gene expression matrices	The raw data from microarray experiments are converted into tables known as gene expression matrices. The rows represent genes and the columns represent experimental conditions. The data in the table are signal intensities, representing relative levels of gene expression.
Grouping expression data	Each gene in a gene expression matrix has an expression profile, that is, the expression measurements over a range of conditions. The analysis of microarray data involves grouping these data on the basis of similar expression profiles. If a predefined classification system is used to group the genes, the analysis is described as supervised. If there is no predefined classification, the analysis is described as unsupervised and is known as clustering.
Clustering methods	Clustering first involves converting the gene expression matrix into a distance matrix, so genes with similar expression profiles can be grouped together. This generally involves calculating the Euclidean distance, the correlation measure based distance or the Pearson linear correlation based distance for each pair of values. Several clustering methods can then be used including hierarchical clustering, *k*-means clustering and the derivation of self-organizing maps.
Feature reduction	A characteristic of microarray data analysis is the large number of features (data points). Clustering and classification algorithms can run more quickly if feature reduction is applied, to remove or amalgamate redundant and noninformative data.
Related topics	Gene expression data (B3) Building phylogenetic trees (G2) Phylogenetics, cladistics and Microarray data: tools and resources (J2) ontology (G1) Analyzing data from 2D-PAGE gels (K1)

Raw data from microarrays

Microarrays are miniature devices comprising a large number of DNA sequences immobilized on a substrate such as a glass microscope slide. The sequences, known as **features**, are arranged as a grid. Arrays are hybridized with a **complex probe**[1] (a population of labeled DNA or RNA molecules, representing a particular cell type or tissue). The intensity of the hybridization signal for each feature corresponds to the amount of that particular molecule in the probe, and this is directly proportional to the level of gene expression in the cell type or tissue from which the probe was prepared. In this way, microarrays can report the relative expression levels of thousands or tens of thousands of genes (Topic B3).

The raw data from microarray experiments thus consist of images from hybridized arrays. The exact nature of the image, however, depends on the **array platform** (the type of array used). As discussed in Topic B3 there are three basic types of array. The first generation of arrays was made by spotting DNA molecules onto **nylon membranes**. This type of array is hybridized with a **radioactive probe**, and signals are detected and quantified using a **phosphorimager**. The spatial resolution of radioactive signals is low, so the features on the array cannot be packed very tightly. Therefore, nylon arrays tend to be large (in the order of $10 \, cm^2$) and are sometimes called **macroarrays** for this reason. Feature density can be increased by using a **colorimetric** label instead of a radioactive label, but the sensitivity is lower.

Most array experiments nowadays are carried out using glass **spotted microarrays** or **high-density oligonucleotide chips**. In both cases, the substrate has minimal autofluorescence so a **fluorescent probe** can be used. The data are acquired by confocal laser scanning of the hybridized array at the appropriate **excitation wavelength** and recording at the appropriate **emission wavelength** (or **channel**). As discussed in Topic B3, a single label is used for oligonucleotide chips, so differential gene expression is detected by hybridizing different probes to duplicate arrays. However, in the case of spotted arrays, two probes can be labeled with different fluorophores and hybridized simultaneously to the same array allowing differential gene expression to be monitored directly.

Data quality

DNA arrays may contain many thousands of features. Therefore, data acquisition and analysis must be automated. The software for initial image processing is normally provided with the scanner (or phosphorimager). This allows the boundaries of individual spots to be determined and the total signal intensity to be measured over the whole spot (this is termed the **signal volume**). Locating spots precisely can be a problem, particularly if there is distortion on the array surface. Therefore, it is often necessary to align the grid manually. This is very important because signal intensities can vary across individual spots and the shape and size of different spots may not be uniform. Most importantly, the signal intensity has to be corrected for **background**, which may be generated by nonspecific hybridization, autofluorescence, dust and other contaminants or poor hybridization technique (e.g. partial dehydration). The background can vary over the array surface, so signal intensities must be normalized for local

---

[1] We use the term *probe* in the traditional sense, that is, referring to the labeled nucleic acid population in solution. The immobilized and unlabeled DNA on the array is thus the *target*. Unfortunately, this terminology is often reversed in the scientific literature, particularly when discussing high-density oligonucleotide chips. The reader should be aware of this when reading papers about microarray experiments!

background values. Correction for background is difficult when the signal intensity for a particular spot is itself very low.

Control features should be included on the array to measure nonspecific hybridization and variable hybridization across arrays. For example, Affymetrix GeneChips incorporate a set of **mismatching oligonucleotides** for each perfect match set to determine nonspecific hybridization. Controls are particularly important where duplicate arrays are being used to study differential gene expression, since variation in array manufacture or experimental protocol can influence the signal intensities on different arrays. The bottom line is that errors and artifacts introduced before or during data acquisition cannot be detected or corrected at a later stage.

Gene expression matrices

Whichever platform is used, the aim of data processing is to convert the hybridization signals into numbers, which can be used to build a **gene expression matrix**. Essentially this matrix can be regarded as a table in which the rows represent genes (the different features on the array) and the columns represent treatments or conditions used in the experiment. For a dual hybridization experiment using a glass microarray, each of the probes represents a different experimental condition. In other cases, a whole series of conditions or treatments may be used, for example representing a series of concentrations of a particular drug, or a series of developmental time points.

Figure 1 is an idealized representation of such a matrix, with nine data points. For any series of experiments, the number of data points in the matrix is $N_C \times N_G$, where N_G is the number of genes (number of features on the array for which data is recorded) and N_C is the number of experimental conditions applied.

Grouping expression data

The interpretation of microarray experiments is carried out by grouping the data according to similar expression profiles. An **expression profile** can be defined, in this context, as the expression measurements of a given gene over a set of

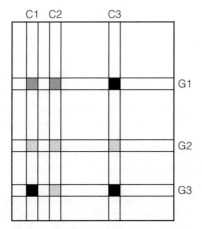

Fig. 1. *Schematic of an idealized expression array, in which the results from three experiments are combined. Three genes (G1, G2 and G3) are labeled on the vertical axis and three experimental conditions (C1, C2 and C3) are labeled on the horizontal axis, giving a total of nine data points. The shading of each data point represents the level of gene expression, with darker colors representing higher expression levels.*

conditions. Essentially this means reading along a row of data in the gene expression matrix. The nature of the problem, which involves multivariate statistics, is shown in *Fig. 1*. In this case, the intensity of shading is used to represent expression levels. If we concentrate on experimental conditions C1 and C2, we can conclude that genes *G1* and *G2* are functionally similar and *G3* appears to be different. However, if we include C3, this suggests a functional link between *G1* and *G3*.

Analysis methods can be described as either supervised or unsupervised. **Supervised methods** are essentially **classification systems**. That is, they incorporate some form of classifier so that expression profiles are assigned to one or more *predefined categories*. For example, the supervised analysis of gene expression profiles from different leukemias allows the samples to be divided into two distinct subtypes: acute myeloid leukemia and acute lymphoblastoid leukemia. **Unsupervised methods** have no inbuilt classifiers, so the number and nature of groups depends only on the algorithm used and the nature of the data themselves. This type of analysis is known as **clustering** and is discussed in more detail below.

Clustering methods

Clustering is a way of simplifying large data sets by partitioning similar data into specific groups (**clusters**). The successive stages of the analysis require a measurement of the distances between the genes in terms of their expression profiles followed by the use of some clustering algorithm. The first step is to convert a gene expression matrix into a **distance matrix**, so that the similarities and differences between data points can be determined. This is often achieved by calculating a value known as the **Euclidean distance**, which is the square root of the sum of the squared differences between any two data points. Alternatively, the **correlation measure-based distance** or the **Pearson linear correlation-based distance** may be used because the Euclidean distance produces misleading clusters if two genes have similar expression profiles but different amplitudes or if two genes have different but very low expression profiles.

Having acquired a distance matrix, the next task is to cluster the data and generate a graph known as a **dendrogram**. Essentially the same process is used to build phylogenetic trees (Topic G2). However, the specific methods used in phylogenetic analysis are not readily applied to expression data, partly because the size of the data sets is much larger and partly because the concept of an 'ancestor' is meaningless, so that a rooted tree is not required. Gene expression clusters are unrooted trees. See Topic G2 for more discussion on this subject.

There are several different clustering strategies that can be applied to expression data, and it is not yet clear which strategy if any is the most suitable. **Hierarchical clustering** methods, similar to those applied in phylogenetic analysis, are the most widely used. Initially each gene defines its own singleton cluster and the algorithm searches for the two most similar clusters. These **neighbors** are then merged into a single cluster and the process repeated. As discussed in more detail in Topic G2, the distance between a new (merged) cluster and any other given gene can be defined as the distance to the nearest of the neighbors in the merged cluster (single linkage), the distance to the farthest of the neighbors (complete linkage) or the distance to the average value of the neighbors (average linkage, with or without weighting). These methods generate dendrograms with different topologies (*Fig. 2*).

Fig. 2. *Four genes (G_1–G_4) are clustered according to their expression profiles in a microarray experiment. Alternative clustering methods lead to trees with different topologies. In this example, tree (a) has been generated using the nearest-neighbors method (single linkage), whereas tree (b) has been generated using the farthest-neighbors method (complete linkage).*

Other, nonhierarchical clustering methods can also be used. One popular method is **k-means clustering**, in which the expected number of clusters is specified at the outset and defined as the parameter k. Initially, the center of each cluster (calculated as the **centroid value**, the weighted equivalent of a center of gravity) is randomly specified. Expression profiles are assigned to a particular cluster according to a distance matrix and then the centroid value is recalculated based on the incorporation of new profiles. Reiteration of this process eventually generates a tree in which groups of genes are assigned to clusters based on similar expression profiles. The advantage over hierarchical methods is that the boundaries between one cluster and another are not defined arbitrarily, but are recalculated in each iteration (see Topic G2). A similar process, refined by the use of neural networks, involves the generation of **Kohonen self-organizing maps (SOMs)**. Centroid values for the clusters are recalculated using not only information from profiles within each cluster, but also from profiles in adjacent clusters.

Feature reduction

Since microarray data sets are so large, classification and clustering can be laborious and demanding in terms of computer resources. It is sometimes possible to use **feature reduction**, where noninformative or redundant data points are removed from the data set, to make the algorithms run more quickly. For example, if two conditions have exactly the same effect on gene expression, these data are redundant and one entire column of the matrix can be eliminated. Similarly, if the expression of a particular gene is the same over a range of conditions, it is neither necessary nor beneficial to use this gene in further analysis because it provides no useful information on differential gene expression. An entire row of the matrix can be removed. Several approaches can be used to automatically select such redundant or noninformative data sets, but a popular method is **principal component analysis** (also called **singular value decomposition**). Redundant data are combined to form a single, composite data set, thus reducing the dimensions of the gene expression matrix and simplifying the analysis. Feature reduction can also be used in supervised analysis methods to reduce the number of features required to classify gene expression profiles correctly (this is sometimes called **cherry picking**). In one method, this can be achieved simply by weighting classification features according to their usefulness and eliminating those that are least informative.

J2 MICROARRAY DATA: TOOLS AND RESOURCES

Key Notes

Microarray data format

Unlike sequence and structural data, there is no international convention for the representation of data from microarray experiments. This is due to the wide variation in experimental design, assay platforms and methodologies. Recently, an initiative to develop a common language for the representation and communication of microarray data has been proposed. Experiments are described in a standard format called MIAME and communicated using a standardized data exchange model and microarray markup language based on XML.

Tools for microarray data analysis

Many software applications are available for the analysis of microarray data and these can be downloaded and installed on local computers. There are also several resources, Expression Profiler being the most widely used, for microarray data analysis over the Internet. Several gene expression databases have been constructed for the storage and dissemination of microarray data. These include the NCBI Gene Expression Omnibus and the EBI ArrayExpress database.

From expression data to pathways

Reconstructing molecular pathways from expression data is a difficult task. One approach is to simulate pathways using a variety of mathematical models and then choose the model that best fits the data. Reverse engineering is a less demanding approach in which models are built on the basis of the observed behavior of molecular pathways. Models using simultaneous differential equations or Boolean networks each suffer from disadvantages, so hybrid models, such as the finite linear state model, are preferred.

Related topics

Gene expression data (B3)
Miscellaneous databases (C4)

Microarray data: analysis methods (J1)
Software downloading and installation (O3)

Microarray data format

Traditionally, bioinformatics has dealt with the analysis of sequence and structural data. There is a standard convention for the presentation of nucleic acid and protein sequences, and atomic coordinates in protein structures, allowing these data to be interpreted unambiguously by scientists around the world. More recently, the scope of bioinformatics has widened to include the analysis of gene and protein expression data. A standard format has been adopted for the representation of two-dimensional gel electrophoresis (2D-PAGE) protein gels (see Topic K1) but there is no similar convention for microarrays, even though microarray experiments produce some of the largest data sets bioinformatics has to deal with. This reflects the different array platforms available (i.e. nylon macroarrays, spotted glass microarrays, high-density oligonucleotide chips) and the large amount of variation in experimental design, hybridization protocols and data-gathering techniques.

Recently, there has been an international effort to develop a common language for the communication of microarray data. Essentially, the requirements for this language are that it should be minimal (i.e. no unnecessary embellishment of the experimental design and protocol) but that is should convey enough information to enable the experiment to be repeated, if necessary. The convention is known as **MIAME (minimum information about a microarray experiment)** and incorporates six elements: overall experimental design, array design (i.e. identification of each spot on each array), probe source and labeling method, hybridization procedures and parameters, measurement procedure (including normalization methods) and control types, values and specifications.

The language for expression data communication has been devised by the **MAGE group (microarray and gene expression group)**. Currently, this involves a data exchange model (**MAGE-OM, microarray gene expression object model**) and a data exchange format (**MAGE-ML, microarray gene expression markup language**). MAGE-OM is modeled using the **Unified Modeling Language (UML)** and MAGE-ML uses **XML (eXtensible Markup Language)**. Information on these developments can be found in the regularly updated **Microarray Gene Expression Database (MGED)**, which is located at the following URL: http://www.mged.org.

Tools for microarray data analysis

Microarray data analysis methods were discussed in Topic J1. Many software applications are available for implementing these methods, and a list of useful resources is shown in *Table 1*. In many cases, the analysis is carried out by installing and running software locally, but other applications can be used over the Internet. The European Bioinformatics Institute (EBI) site provides one of the most user-friendly World Wide Web (WWW)-based microarray analysis tools, which is called **Expression Profiler** (EP; http://ep.ebi.ac.uk/EP/). This is a suite of programs designed for the analysis and integration of sequence and expression data. The most relevant program for microarray analysis is **EPCLUST**, which allows data to be uploaded from source or accessed from publicly available files and grouped using a variety of alternative methods for distance measurement and clustering. A useful tutorial to the EP suite of programs can be found at the following URL: http://ep.ebi.ac.uk/EP/doc/Tutorial.html.

From expression data to pathways

Functional information about genes can often help in the reconstruction of molecular pathways and networks, such as metabolic pathways, signal transduction cascades and regulatory hierarchies (Topic L1). For example, the mutant phenotypes of genes acting in the same molecular pathway are often very similar, and pathways can also be deduced from information about protein-protein interactions. Genes in the same pathway may also have similar expression profiles. However, the problem of reconstructing molecular pathways from expression data is difficult and has not been fully solved.

The first step in the reconstruction of a molecular pathway is to build a mathematical model. Models can be developed that predict the behavior of a pathway under different operational circumstances, and then tested by observing the behavior of the molecular pathway itself. This approach is known as **simulation**. For any given number of genes, there is a finite (if very large) number of possible models. Exhaustive simulation would involve the testing of

Table 1. Internet resources for microarray expression analysis. The first two sites are very comprehensive and contain hundreds of links to databases, software and other resources. Two web-based suites of analysis programs are also listed as well as some databases that store microarray and other gene expression data

URL	Product(s)	Comments
Sites with extensive links to microarray analysis software and resources		
http://genome-www4.stanford.edu/MicroArray/SMD/restech.html	Cluster, XCluster, SAM, Scanalyze, many others	Extensive list of software resources from Stanford University and other sources, both downloadable and WWW-based
http://ihome.cuhk.edu.hk/~b400559/arraysoft.html	Cluster, Cleaver, GeneSpring, Genesis, many others	Comprehensive list of downloadable and WWW-based software of microarray analysis and data mining, plus links to gene expression databases
WWW-based microarray data analysis		
http://ep.ebi.ac.uk/EP/	Expression profiler	Very powerful suite of programs from the EBI for analysis and clustering of expression data.
http://bioinfo.cnio.es/dnarray/analysis/	DNA arrays analysis tools	A suite of programs from the National Spanish Cancer Center (CNIO) including two-sample correlation plot, hierarchical clustering, SOM, neural network and tree viewers
Microarray databases		
http://www.ncbi.nlm.nih.gov/geo/	National Center for Biotechnology Information (NCBI)	GEO (Gene Expression Omnibus) GEO is a gene expression and hybridization array database, which can be searched by accession number, through the contents page, or through the Entrez ProbeSet search interface (Topic D1)
http://www.ebi.ac.uk/microarray/ArrayExpress/arrayexpress.html	ArrayExpress	EBI microarray gene expression database. Developed by MGED and supports MIAME (see main text)
http://www.ncgr.org/genex/	GeneX	The GeneX gene expression database is an integrated tool set for the analysis and comparison of microarray data

each of these possible models and finding the one that was most compatible with the observed behavior of the pathway. This is suitable for small networks, but as the number of genes increases, the number of possible networks becomes too large to work with. Alternatively, models can be created that are consistent with the observed behavior of the molecular pathway. This is known as **reverse engineering** and is a subject of intense current research.

There are two well-studied ways of representing a molecular pathway or network. The classical biochemical representation involves the use of **simultaneous differential equations,** so that for three substances s_1, s_2 and s_3 (e.g. substrate, intermediate and product in a metabolic pathway) there are three rates of change (t is time):

$$ds_1/dt = f_1(s_1, s_2, s_3)$$
$$ds_2/dt = f_2(s_1, s_2, s_3)$$
$$ds_3/dt = f_3(s_1, s_2, s_3)$$

The functions, f_1, f_2 and f_3 describe the behavior of the pathway and these are usually complicated expressions, including hyperbolic terms such as the Michaelis–Menten equation.

The alternative representation of a molecular pathway is a **Boolean network**. This is a system in which the state of the pathway depends on its previous state, and any change is dependent on an explicit set of rules. As an example, consider the same three substances – s_1, s_2 and s_3 – discussed above. We can implement a set of rules by which the original state of the system (the input, s_1, s_2 and s_3) can be changed into a new state (the output, s_1', s_2' and s_3'). The rules are stated below and *Table 2* shows the permitted output values for any given input.

- Input and output values can be either 0 or 1
- $s_1' = s_2$ (s_1' is the same as s_2)
- $s_3' = \neg s_1$ (s_3' is the opposite of s_1)
- $s_2' = s_1 \ \& \ s_3$ (s_2' is 0 unless both s_3' and s_1 are 1)

Neither simultaneous differential equations nor Boolean networks adequately model real molecular pathways. The problem with simultaneous differential equations is that they do not easily incorporate discrete aspects of the system, such as transcription factor binding and dissociation. Conversely, Boolean networks do not easily incorporate aspects of the system that change continuously, such as the concentrations of metabolic intermediates, mRNA molecules or proteins. For this reason, hybrid models are preferred. An example of such a model is the **finite linear state model**, which incorporates both Boolean logic and aspects of continuous change. The outline of the model is shown in *Fig. 1*. The whole system represents a gene. B1, B2 and B3 are binding sites in the gene's promoter, which may be either occupied (1) or unoccupied (0). The control function (F1) is determined by a set of rules, which might be similar to those used in *Table 2*. The output (a value of 0 or 1) determines whether the gene remains inactive (0) or transcriptionally active (1). The linear output will be in the form of an increase in concentration (in this case of the gene's mRNA).

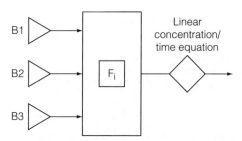

Input   Control function   Linear output

Fig. 1. Outline of a gene as represented by the finite linear state model. B1, B2 and B3 are inputs and represent transcription factor binding sites. They are described in terms of discrete states: 1 (occupied) or 0 (unoccupied). The control function F1 is a set of Boolean rules, perhaps similar to those shown in Table 2, that determines how the input states govern gene expression. The output value is also discrete: 1 (transcriptionally active) or 0 (transcriptionally silent). The linear output, which can be positive, negative or zero, reflects the rate of increase or decrease in the amount of mRNA.

Table 2. Permitted output values (s₁′, s₂′ and s₃′) given all possible input values (s₁, s₂ and s₃). The rules governing this system are shown in the main text. Data modified from Akutsu T. et al., Identification of genetic networks from a small number of gene expression patterns under the Boolean network model, Proceedings of the Pacific Symposium on Biocomputing, 1999, pp. 17–28

s_1	s_2	s_3	s_1'	s_2'	s_3'
0	0	0	0	0	1
0	0	1	0	0	1
0	1	0	1	0	1
0	1	1	1	0	1
1	0	0	0	0	0
1	0	1	0	1	0
1	1	0	1	0	0
1	1	1	1	1	0

Figure 2 shows how a model can be used to describe a real regulatory network, the choice between lysis and lysogeny faced by bacteriophage λ when it infects *Escherichia coli*. Initially, the bacteriophage is committed to neither pathway, but a series of molecular events depending on the growth conditions of the host cell lead to the expression of either *cro* or *CI*. Each of these proteins can bind to the operators (O_R1, O_R2 and O_R3), which are shown as inputs in the model. If *cro* is expressed, the Cro protein binds to the operator sites and prevents transcription from the promoters PR and PM so that the *CI* gene is not transcribed. This establishes lysis. If *CI* is expressed, the CI protein binds to the operator sites, preventing transcription from promoter PR (which prevents the transcription of *cro*) but allowing continued transcription of *CI* through the maintenance promoter PM. This establishes lysogeny. Hybrid methods are adequate for the modeling of such pathways whose behavior is well understood. However, models of unknown pathways based on expression data alone must be rigorously tested.

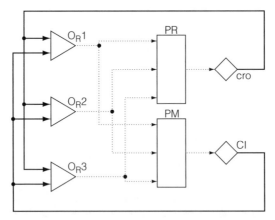

Fig. 2. Model of a real regulatory network, the choice between lysis and lysogeny in bacteriophage λ, represented by inputs, control functions and outputs. See text for details.

J3 SEQUENCE SAMPLING AND SAGE

Key Notes

Sequence sampling data analysis	Differential gene expression can be investigated by sampling random clones from different cDNA libraries, or by sampling EST data, which is obtained by single-pass sequencing of randomly picked cDNA clones and deposited in public or propriety databases. Thousands of sequences have to be sampled for such analysis to be statistically significant, even in the case of moderately abundant mRNAs.
SAGE	SAGE is a sequence sampling technique in which very short sequence tags (9–15 nt) are joined into long concatamers. The size of the SAGE tag is optimal for high-throughput analysis but genes can still be identified unambiguously. A concatamer may contain more than 50 tags, and each SAGE sequence is thus equivalent to more than 50 independent cDNA sequencing experiments. SAGE is therefore appropriate for the analysis of rare mRNAs.
Related topics:	DNA, RNA and protein sequencing (B1) Gene expression data (B3) Annotated sequence databases (C2)

Sequence sampling data analysis

As discussed in Topic B3, gene expression analysis can be carried out by the **direct sampling** of sequence from cDNA (copy DNA) libraries. The way this is achieved is to perform **single-pass sequencing** (one sequence reaction only) of randomly picked clones from cDNA libraries, thus generating short (200–300 bp) **expressed sequence tags (ESTs)**. These sequences, which can be generated very rapidly, are deposited in databases. The major public database, a subsidiary of GenBank, is **dbEST** (Topic C2). Other EST databases have been established by companies such as Incyte Pharmaceuticals Inc. If the raw sequence data are deposited directly into such databases, each EST represents an independent sampling of the original clone library. Therefore, abundant clones are likely to be represented by many ESTs and rare clones are likely to be represented by only a few ESTs. Tools are available over the Internet for the comparison of sequence samples from many different cDNA libraries. These tools include **Virtual Northern** and **xProfiler**, both of which are available from the National Center for Biotechnology Information (NCBI) home page (http://www.ncbi.nlm.nih.gov).

Using current statistical methods, direct sequence sampling can be applied only to abundant transcripts. This is because a sequence must be sampled at least three times for the results of the comparison to be statistically significant. In the case of a rare transcript, which may be present for example as one copy per 50 000 mRNA molecules in the cell, 150 000 sequencing reactions would have to be carried out to generate sufficient data, and this would be prohibitively expensive.

SAGE

SAGE (serial analysis of gene expression) is a sequence sampling technique that circumvents the problems associated with random clone sequencing by sampling many different cDNAs in one experiment. This is achieved by using very short ESTs (9–15 nt) and joining them into a long concatamer prior to sequencing. The ESTs are just long enough to allow accurate and unambiguous gene identification.

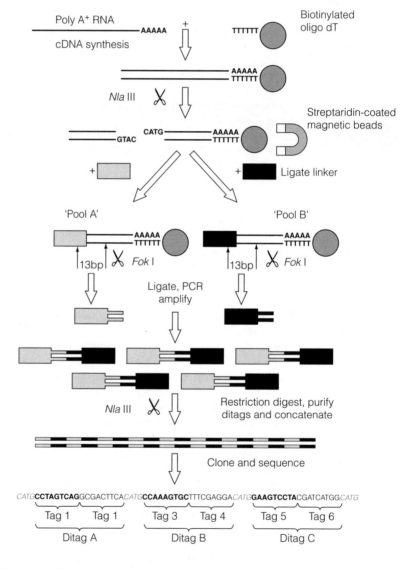

Fig. 1. Simplified outline method for serial analysis of gene expression. NlaIII is a frequent cutting restriction enzyme used initially to generate the 3' cDNA fragments and provide the overhang for linker ligation, and later to remove the linkers prior to concatamerization of the ditags. FokI is a type IIs restriction enzyme with a recognition site in the linker that generates the SAGE tags by cutting the DNA a few bases downstream. From Pennington S.R. and Dunn M.J., Proteomics: From Protein Sequence to Function, © BIOS Scientific Publishers Limited, 2001.

The SAGE method appears quite complex, but can be summarized as follows (*Fig. 1*):

- An mRNA population is reverse transcribed to generate cDNA, using a biotinylated oligo-dT primer.
- The cDNA is digested with a restriction enzyme that cleaves very frequently, such as *Nla*III, leaving 3' fragments of a few hundred base pairs labeled with biotin.
- The 3' end of each cDNA is then captured by affinity to streptavidin, and the other (unlabelled) fragments are discarded.
- The cDNA is divided into two pools, and each pool is ligated to a different **linker** (a double-stranded oligonucleotide with a specific overhang, in this case matching the overhang generated by *Nla*III). The linker carries an internal recognition site for a **type IIs restriction enzyme** such as *Fok*I. These enzymes have the unusual property of cutting outside their recognition site, but a specific number of base pairs away.
- Cleavage with the type IIs enzyme generates a very short sequence tag attached to the linker, and this is known as a **SAGE tag**.
- The biotinylated fragments are discarded, and SAGE tags with different linkers are ligated 'tail-to-tail' to generate dimers called **ditags**.
- The ditags are amplified by the polymerase chain reaction (PCR) using primers that anneal to the linkers.
- The linkers are removed using *Nla*III and the ditags are randomly joined together in a long concatamer.
- The concatamer is sequenced.
- Sequence analysis identifies the tags and thus the associated mRNAs. The number of times each tag appears in the concatamer is indicative of the mRNA's abundance.

As with the random cDNA sampling approach discussed above, comparative SAGE results must be tested for statistical significance. However, due to the large number of tags that can be 'read' in any experiment, the SAGE technique is useful for mRNAs that have an abundance of less than 1 in 10 000. Tools for SAGE data processing are available on the Internet, as well as databases of SAGE experiments. A list of useful resources is provided in *Table 1*. One disadvantage of SAGE is that sequencing errors have a profound effect since one base miscall can result in the misidentification of a SAGE tag. This is not the case in the random clone sequencing approach because in this case the ESTs are much longer.

Table 1. Starting points for SAGE analysis on the Internet

Resource	URL
John Hopkins SAGE site. Includes protocols, access to SAGE data and an extensive bibliography	http://www.sagenet.org/
NCBI SAGE site. Includes tools for data analysis, access to SAGE data, and library of tags and ditags	http://www.ncbi.nlm.nih.gov/SAGE
Saccharomyces genome database SAGE query site	http://genome-www.stanford.edu/cgi-bin/SGD/SAGE/querySAGE
A useful SAGE site run by Genzyme Molecular Oncology Inc., which owns the license for commercial distribution of SAGE technology	http://www.genzymemolecularoncology.com/sage/

K1 ANALYZING DATA FROM 2D-PAGE GELS

Key Notes

Raw data from 2D-PAGE gels

2D-PAGE is a protein separation technique that allows the resolution of thousands of proteins on a single gel, on the basis of charge and mass. Separated proteins appear as spots, the nature and distribution of which constitute a protein fingerprint of any sample.

Data processing

Data extraction from 2D-PAGE gels involves staining (to reveal the position of individual protein spots), scanning (to obtain a digital image) and then spot detection and quantitation. The quality of the image, in terms of spatial and densitometric resolution, is an important factor in accurate spot measurement. A number of algorithms are used to resolve complex overlapping spots and assemble a final spot list.

Gel matching

To study differential protein expression, a series of 2D-PAGE gels must be compared. However, minute inconsistencies in gel structure and electrophoretic conditions make it impossible to exactly replicate any experiment. Sophisticated algorithms are required to follow individual spots through a series of gels, a process known as gel matching. MELANIE II is a widely used gel-matching software application.

Protein expression matrices

Differential protein expression data are assembled into a protein expression matrix. This can be used to find distances between particular proteins or treatments, leading to classification or clustering of proteins according to similar expression profiles.

2D-PAGE databases

Data from 2D-PAGE experiments are deposited in dedicated 2D-PAGE databases containing digital gel images with links from individual protein spots to useful annotations. Internet 2D-PAGE databases are indexed at the ExPASy WORLD-2DPAGE. These allow 2D-PAGE data to be shared with scientists around the world, and comparisons between gels can be carried out using Java applets such as Flicker or CAROL.

Related topics

Gene expression data (B3)
Annotated sequence databases (C2)
Microarray data: analysis methods (J1)

Microarray data: tools and resources (J2)
Analyzing protein mass spectrometry data (K2)

Raw data from 2D-PAGE gels

Two-dimensional polyacrylamide gel electrophoresis (2D-PAGE) is a method used to separate proteins according to charge and mass (Topic B3). The resolution of the technique is such that thousands of proteins can be distinguished, providing a diagnostic **protein fingerprint** of any particular sample. After the gel has been run, it is stained to reveal the position of individual proteins. These appear as **spots** of varying size, shape and intensity. The role of bioinformatics in protein gel analysis is the extraction of useful information from the positions and intensities of the protein spots.

2D-PAGE data can be used to derive general information on protein expression profiles without any knowledge of which specific proteins are actually present on the gel. However, the most powerful approach to 2D-PAGE experiments is to couple the expression analysis with protein annotation by mass spectrometry, and this is discussed in Topic K2.

Data processing

Data are extracted from 2D-PAGE gels in several stages. First, the stained protein gel is scanned to obtain a digital image. Individual protein spots are then detected and quantified, and the intensity of the signal for each spot is corrected for local background.

The quality of the digital image is important since its **spatial resolution** determines how accurately protein spot sizes are recorded and its **densitometric resolution** determines how accurately the intensity of each spot is recorded. Protein spot measurement involves a special set of problems that are not found with microarray data (Topic J1). The features on a microarray are arranged in a precise grid and the signals tend to be regular, discrete and nonoverlapping. Conversely, the signals in a protein gel are irregularly distributed, the spots vary widely in morphology, and it is often the case that spots join together in clumps or lines that are difficult to resolve. Several algorithms are available which address these issues and generally these are based on either **Gaussian fitting** or **Laplacian of Gaussian (LOG) spot detection.** Spots whose morphology deviates from a single Gaussian shape can thus be interpreted using a model of overlapping shapes. A simpler approach is **line and chain analysis**, in which columns of pixels from the digital image are scanned for peaks in signal density. This process is repeated for adjacent pixel columns allowing the algorithm to identify the centers of spots and their overall signal intensity (**signal volume**). A further approach is known as **watershed transformation (WST).** In this method, pixel intensities are viewed as a topographical map so that hills and valleys can be identified. This is useful for separating clusters, chains and shouldered spots (small spots overlapping with larger ones) and also for merging regions of a single spot. The output of each method is a **spot list**, in which each individual protein spot is identified by the x and y coordinates of its center.

Gel matching

An important application of 2D-PAGE is the analysis of **differential protein expression**. This can be used, for example, to look for proteins that are induced or repressed by particular treatments or drugs, to look for proteins associated with disease states, or to look at changes in protein expression during development. Differential protein expression can only be studied by running a series of 2D-PAGE gels with alternative samples, and searching for novel spots or spots whose intensity changes significantly in different gels. Note that any observed changes, for example the presence of a novel protein spot, may not reflect the *synthesis* of a new protein. Instead, the new sport may reflect the **post-translational modification** of an existing protein, for example phosphorylation or glycosylation. Such modifications can radically alter the mass or charge of a protein, leading to different migration behavior during electrophoresis.

Due to minute variations and inconsistencies in the chemical and physical properties of electrophoretic gels, it is impossible to exactly reproduce the conditions of any one electrophoresis experiment. This means that, even in a series of gels using exactly the same sample and electrophoresis parameters, the positions of individual protein spots are never the same. In order to compare serial gels and identify novel protein spots or spots that vary in intensity, it is necessary to use **gel matching**. Generally this involves establishing the positions of several unambiguous **landmark spots** and then using algorithms to match the positions of the remaining spots (*Fig. 1*). To bring the spots from two gels into register, simple image manipulations such as stretching and rotating can be carried out. This may be assisted by incorporating spot intensities and the (physical) distances between neighboring spots as variables in the algorithm. For example, an approach known as **propagation** involves determining the distance between a given landmark spot and all neighbors. If matches are found on other gels, the neighboring spots can be used as new landmarks, and the process reiterated. An implementation of such algorithms is available as the program **MELANIE II**. This is available to download for use on Windows-based computers (www.expasy.ch/ch2d/melanie) but cannot be used online.

Protein expression matrices

Once protein expression data have been recorded they are built into a **protein expression matrix**. This is similar in principle to the gene expression matrix discussed in Topic J1. Spots, representing proteins, are arranged in rows, while experimental treatments are listed in columns. The data points in the matrix represent signal intensities for each spot. Comparison of values along a given row can identify proteins whose expression levels change according to different treatments. As with gene expression data, multivariate statistical analysis can group data by similar expression profiles, which can be used both for classification purposes or clustering (Topic J1).

Image analysis of two-dimensional gels

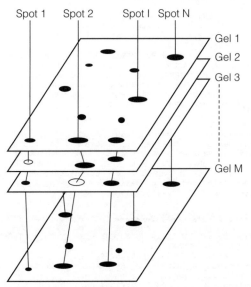

Fig. 1. Principle of gel matching. Individual spots are compared and matched across a series of 2D-PAGE gels. From Pennington S.R. and Dunn M.J., Proteomics: From Protein Sequence to Function © BIOS Scientific Publishers Limited, 2001.

2D-PAGE
databases

The results from 2D-PAGE experiments are generally stored in **2D-PAGE data-bases**, which use gel images as a basis for protein annotation. Digital images from 2D-PAGE gels are presented, and each protein spot acts as a link to further information such as protein name, molecular mass, pI value, annotations from the SWISS-PROT database, bibliographic references and, if appropriate, graphs showing how spot intensities vary over a series of gels. Some 2D-PAGE data-bases are free standing and may come packaged with gel analysis software. For example, the analysis suite **PDQUEST** incorporates its own database, which may be installed locally and used to store the 2D-PAGE experimental data from a single laboratory. There are also free software packages available that can be used to set up 2D-PAGE databases, such as **Make2ddb**.

The construction of 2D-PAGE databases on the World Wide Web (WWW) allows 2D-PAGE data to be shared and compared over the Internet. There are over 30 such databases currently available, many showing 2D-PAGE data either for specific organisms or specific systems. For example, the HEART-2DPAGE contains 2D-PAGE data related to heart development, physiology and disease. All these pages are indexed by the ExPASy server on the **WORLD-2DPAGE** (http://www.expasy.ch/ch2d/2d-index.html). The format of data presentation is standardized. Images of 2D-PAGE gels are presented overlain with a grid showing pI values and molecular masses. This allows each protein spot to be identified on the basis of its physical and chemical properties. Individual spots can be clicked, linking the user to an annotation page. On many gels, most of the spots are not annotated, and those with annotations are highlighted in some manner. For example, on the SWISS-2DE database each annotated protein spot is marked with a red cross. Annotations can be viewed as an original SWISS-PROT file, an example

```
ID    ACTB_HUMAN; STANDARD; 2DG.
AC    P02570;
DT    01-AUG-1993 (Rel. 00, Created)
DT    01-DEC-2000 (Rel. 13, Last update)
DE    Actin, cytoplasmic 1 (Beta-actin).
GN    ACTB.
OS    Homo sapiens (Human).
OC    Eukaryota; Metazoa; Chordata; Craniata; Vertebrata; Euteleostomi;
OC    Mammalia; Eutheria; Primates; Catarrhini; Hominidae; Homo.
OX    NCBI_TaxID=9606;
MT    CSF_HUMAN, ELC_HUMAN, HEPG2_HUMAN, HEPG2SP_HUMAN, LIVER_HUMAN,
MT    LYMPHOMA_HUMAN, PLASMA_HUMAN, PLATELET_HUMAN, RBC_HUMAN, U937_HUMAN,
MT    KIDNEY_HUMAN, HL60_HUMAN, CEC_HUMAN, DLD1_HUMAN.
IM    CSF_HUMAN, ELC_HUMAN, HEPG2_HUMAN, HEPG2SP_HUMAN, LIVER_HUMAN,
IM    LYMPHOMA_HUMAN, PLASMA_HUMAN, PLATELET_HUMAN, RBC_HUMAN, U937_HUMAN,
IM    KIDNEY_HUMAN, HL60_HUMAN, CEC_HUMAN, DLD1_HUMAN.
RN    [1]
RP    MAPPING ON GEL.
RX    MEDLINE=93162045; PubMed=1286669; [NCBI, ExPASy, EBI, Israel, Japan]
RA    Hochstrasser D.F., Frutiger S., Paquet N., Bairoch A., Ravier F.,
RA    Pasquali C., Sanchez J.-C., Tissot J.-D., Bjellqvist B., Vargas R.,
RA    Appel R.D., Hughes G.J.;
RT    "Human liver protein map: a reference database established by
RT    microsequencing and gel comparison.";
RL    Electrophoresis 13:992-1001(1992).
RN    [2]

.....Several more literature references follow.....

CC    -!- SUBUNIT: SINGLE CHAIN WHICH CAN BIND UP TO 4 OTHER CHAINS.
2D    -!- MASTER: CSF_HUMAN;
2D    -!-   PI/MW: SPOT 2D-000C1S=5.24/44747;
2D    -!-   MAPPING: MATCHING WITH THE PLASMA MASTER GEL [2].
2D    -!- MASTER: ELC_HUMAN;
2D    -!-   PI/MW: SPOT 2D-000ED0=5.21/41208;
2D    -!-   PI/MW: SPOT 2D-000ED7=5.12/41300;
2D    -!-   MAPPING: MATCHING WITH THE LIVER MASTER GEL [2].
2D    -!- MASTER: HEPG2_HUMAN;
```

```
2D   -!-   PI/MW: SPOT 2D-00030B=5.15/41700;
2D   -!-   PI/MW: SPOT 2D-00030Z=5.09/41700;
2D   -!-   PI/MW: SPOT 2D-00031Z=5.23/41272;
2D   -!-   MAPPING: MATCHING WITH THE LIVER MASTER GEL [2].
2D   -!- MASTER: HEPG2SP_HUMAN;
2D   -!-   PI/MW: SPOT 2D-000952=5.27/38525;
2D   -!-   PI/MW: SPOT 2D-000954=5.16/38400;
2D   -!-   PI/MW: SPOT 2D-000955=5.22/38223;
2D   -!-   PI/MW: SPOT 2D-000959=5.11/38223;
2D   -!-   MAPPING: MATCHING WITH THE PLASMA MASTER GEL [2].
2D   -!- MASTER: LIVER_HUMAN;
2D   -!-   PI/MW: SPOT 2D-0000WF=5.26/41839;
2D   -!-   PI/MW: SPOT 2D-0000WN=5.19/41722;
2D   -!-   PI/MW: SPOT 2D-0000WO=5.22/41605;
2D   -!-   MAPPING: MATCHING WITH A PLASMA GEL [1].
2D   -!- MASTER: LYMPHOMA_HUMAN;
2D   -!-   PI/MW: SPOT 2D-0007PE=5.26/41898;
2D   -!-   PI/MW: SPOT 2D-0007PQ=5.15/42194;
2D   -!-   MAPPING: MATCHING WITH THE LIVER MASTER GEL [2].
2D   -!- MASTER: PLASMA_HUMAN;
2D   -!-   PI/MW: SPOT 2D-00050N=5.28/43590;
2D   -!-   PI/MW: SPOT 2D-00050Q=5.24/43244;
2D   -!-   MAPPING: IMMUNOBLOTTING [3].
2D   -!-   NORMAL LEVEL: PLATELET CONTAMINATION.
2D   -!- MASTER: PLATELET_HUMAN;
2D   -!-   PI/MW: SPOT 2D-000FWX=5.27/41946;
2D   -!-   PI/MW: SPOT 2D-000FZ5=5.18/41400;

2D   -!-   PI/MW: SPOT 2D-000FZM=5.06/41400;
2D   -!-   PI/MW: SPOT 2D-000FZN=5.30/41400;
2D   -!-   MAPPING: MATCHING WITH RBC AND LIVER MASTERS [5].
2D   -!- MASTER: RBC_HUMAN;
2D   -!-   PI/MW: SPOT 2D-00064D=5.20/42104;
2D   -!-   PI/MW: SPOT 2D-000640=5.14/42209;
2D   -!-   PI/MW: SPOT 2D-00064W=5.27/42104;
2D   -!-   MAPPING: IMMUNOBLOTTING [3] AND MATCHING [4].
2D   -!- MASTER: U937_HUMAN;
2D   -!-   PI/MW: SPOT 2D-000CXH=5.23/41807;
2D   -!-   MAPPING: MATCHING WITH THE LIVER MASTER GEL [2].
2D   -!- MASTER: KIDNEY_HUMAN;
2D   -!-   PI/MW: SPOT 2D-000N5T=5.18/41212;
2D   -!-   PI/MW: SPOT 2D-000N5U=5.25/41503;
2D   -!-   PI/MW: SPOT 2D-000N67=5.14/41406;
2D   -!-   PI/MW: SPOT 2D-000N6J=5.12/41309;
2D   -!-   MAPPING: MATCHING WITH THE LIVER MASTER GEL AND IMMUNODETECTION
2D         [6].
2D   -!- MASTER: HL60_HUMAN;
2D   -!-   PI/MW: SPOT 2D-000YZK=5.25/41925;
2D   -!-   PI/MW: SPOT 2D-000Z0A=5.15/41390;
2D   -!-   PI/MW: SPOT 2D-000Z0G=5.08/41497;
2D   -!-   MAPPING: MATCHING WITH THE LIVER MASTER GEL [7][8].
2D   -!- MASTER: CEC_HUMAN;
2D   -!-   PI/MW: SPOT 2D-000TWW=5.05/41396;
2D   -!-   PI/MW: SPOT 2D-000TWX=5.11/41497;
2D   -!-   MAPPING: MATCHING WITH THE LIVER MASTER GEL [9].
2D   -!- MASTER: DLD1_HUMAN;
2D   -!-   PI/MW: SPOT 2D-001E5J=5.15/41545;
2D   -!-   PI/MW: SPOT 2D-001E5O=5.19/41545;
2D   -!-   PI/MW: SPOT 2D-001E6E=5.10/41391;
2D   -!-   PEPTIDE MASSES: SPOT 2D-001E5J: 976.504; 1132.57; 1198.7;
2D         1516.72; 1790.89; 1954.05; 2231.02; TRYPSIN.
2D   -!-   PEPTIDE MASSES: SPOT 2D-001E5O: 976.464; 1132.53; 1198.7;
2D         1516.7; 1790.87; 1954.05; 2231.07; TRYPSIN.
2D   -!-   PEPTIDE MASSES: SPOT 2D-001E6E: 976.513; 1132.57; 1516.73;
2D         1790.89; 1954.04; 2231.05; TRYPSIN.
2D   -!-   MAPPING: MASS FINGERPRINTING [10].
CC   ---------------------------------------------------------------------
CC   This SWISS-2DPAGE entry is copyright the Swiss Institute of Bioinformatics.
CC   There are no restrictions on its use by non-profit institutions as long as
CC   its content is in no way modified and this statement is not removed. Usage
CC   by and for commercial entities requires a license agreement (See
CC   http://www.isb-sib.ch/announce/ or send an email to license@isb-sib.ch).
CC   ---------------------------------------------------------------------
DR   SWISS-PROT; P02570; ACTB_HUMAN.
DR   Siena-2DPAGE; P02570; ACTB_HUMAN.
//
```

Fig. 2. Example SWISS-2DPAGE entry (for human actin B). Some material has been deleted for brevity as shown. Note the extensive 2D data, which includes information from 2D-PAGE experiments and mass spectrometry.

of which is shown in *Fig. 2*. Most of the fields are self-explanatory (see Topic C2), but note that the 2D field may include data on both 2D-PAGE electrophoresis and peptide mass fingerprinting (Topic K2) where such experiments have been carried out.

Another WWW resource, **2D Web Gel (2DWG)** can be found at the following URL: http://www-lecb.ncifcrf.gov/2dwgDB/. This is a catalog of some of the 2D gel images found on the WWW, with associated search facilities so that images can be abstracted by key words. 2DWG represents a very good introduction to the world of 2D gels. The images are stored in formats such as GIF that Internet browsers can handle and there is an associated Java applet, called **Flicker**, to facilitate the comparison of two gels. The basis of flickering is that the images of two gels can be rapidly alternated to identify matching spots. Another useful applet is **CAROL** (http://gelmatching.inf.fu-berlin.de/Carol.html) which uses point pattern matching to compare any two gel images over the Internet.

K2 ANALYZING PROTEIN MASS SPECTROMETRY DATA

Key Notes

Raw data from mass spectrometry

The raw data from mass spectrometry experiments are the mass/charge (m/z) ratios of ions in a vacuum. These are used to determine accurate molecular masses. The masses can be used in peptide mass fingerprinting or fragment ion searching to find correlations in protein databases. Alternatively, peptide ladders can be generated and used to determine protein sequences *de novo*.

Virtual digests

Virtual digests are theoretical protein cleavage reactions performed by computers based on known protein sequences and the known specificity of a cleavage agent such as an endoproteinase. Although many different polypeptides can generate the same peptide digest pattern, in practice a correlation between the masses of two or more peptides produced from the same protein and the theoretical peptides produced in a virtual digest provides very strong evidence for a database match.

Dual digests

Dual digests, carried out on the same protein either separately or sequentially, can provide extra data to correlate experimentally determined molecular masses with less robust data resources such as dbEST. Alternatively, single digests can be carried out before and after protein modification, or ragged termini can be generated from proteins with clustered arginine and lysine residues, providing the masses of multiple fragments to use as database search terms.

Database search tools

Algorithms for database searching may attempt to match the experimentally determined mass of a peptide or peptide fragment to masses predicted from sequence database entries. The program SEQUEST works on this principle. Alternatively, the amino acid composition of a particular peptide or peptide fragment can be predicted from its mass. The order of amino acids cannot be predicted, so all permitted permutations are used as a database search query. The program Lutkefisk works on this principle.

Limitations of MS analysis

The failure of MS data to elicit a high-confidence hit on a sequence database may not always reflect the absence of that protein from the database. In some cases, it may reflect the presence of unknown or unanticipated post-translational modifications, or it may be caused by nonspecific proteolysis or contaminating proteins. Imperfect matches may be generated if the experimental protein itself is absent from the database but a close homolog, with a related sequence, is present.

Related topics

Sequencing of DNA, RNA and proteins (B1)

Annotated sequence databases (C2)
Analyzing data from 2D-PAGE gels (K1)

Raw data from mass spectrometry

Mass spectrometry (MS) is a method for accurately determining the mass/charge ratio (m/z) of ions in a vacuum, thus allowing the precise determination of molecular masses. These raw data can be used in three different approaches for the identification of proteins. In peptide-mass fingerprinting, a protein is digested with a specific cleavage agent (usually the enzyme trypsin) and the masses of each of the resulting peptides are determined. These masses are then used in correlative database searching to identify the protein. In fragment ion searching, peptide fragments are generated as above and then fragmented in a collision cell between two quadrupole mass analyzers (this is called tandem MS, often abbreviated to MS/MS) or by a process known as post-source decay, which occurs when matrix-assisted laser desorption/ ionization (MALDI) MS is used at a higher acceleration voltage than usual (see Topic B1 for further explanation of these techniques). The resulting fragment ions are short peptide fragments. The molecular masses of such fragments can be used to search not only protein databases, but also other sequence repositories including dbEST (Topics B1 and C2). Both these approaches require that the protein in question has been identified and its sequence deposited. Where this is not the case, *de novo* sequencing can be carried out. In this method, peptide ladders are generated either by sequential chemical degradation of terminal amino acid residues or by separating the fragment ions generated as above into nested sets. Mass differences between sequential fragments correspond to the known masses of individual amino acids, thus allowing protein sequences to be deduced without any correlative information from sequence databases.

Virtual digests

It would be impossible to link peptide mass data to proteins in sequence databases such as SWISS-PROT without first knowing the *expected* peptides from such proteins. This information can be obtained by carrying out virtual digests, that is, theoretical digests based on the known protein sequence and the known specificity of the cleavage agent used. Cleavage agents with high specificity are most suitable and the endoproteinase trypsin, which cleaves a polypeptide chain after each basic amino acid (lysine or arginine) providing the next residue is not a proline, is the most widely used. Given a protein of known sequence, the tryptic peptides can be predicted. For example, a protein with the sequence shown below

MCLTAKGAATCSATFRYLIFALSLATKPACALLASALLARACATTAVA

would generate the following tryptic peptides:

MCLTAK GAATCSATFR YLIFALSLATKPACALLASALLAR ACATTAVA.

The four peptides provide four theoretical molecular masses. Correlation between these theoretical masses and the actual masses obtained in a mass spectrometry experiment would provide very convincing evidence for a database match. Of course the same molecular masses could be obtained in many other ways. Each of the four peptides has a potentially very large number of anagrams (same amino acids in a different order), which would all have the same molecular mass. Theoretically, this places limitations on the usefulness of the technique but in practical terms the chances of a series of peptides from the same protein generating spurious matches to the same deposited protein sequence because of permutations in the order of amino acids are very small indeed. Generally, correlation between two or more peptides is taken to be unambiguous confirmation of a database match. Another theoretical limitation is that the amino acids leucine and isoleucine have the same molecular masses. This provides a small technical problem in *de novo* sequencing, but for peptide mass fingerprinting and fragment ion fingerprinting it does not have a practical impact.

Dual digests

Peptide mass fingerprinting allows the rapid identification of proteins if they are already represented in databases such as SWISS-PROT. Where this is not the case, both peptide mass fingerprinting and fragment ion searching can be used to match mass spectrometry data to other sequence databases, including the expressed sequence tag (EST) database dbEST. The problem with this approach is that the databases contain a large amount of irrelevant data (e.g. noncoding sequence and other 'noise'), which can reduce the efficiency of the search. More confidence can be placed in any results if **dual digests** are carried out (i.e. combining the information from two protease digests using enzymes with different specificities, such as trypsin and endoproteinase LysC). Another approach is to carry out a single digest with the protein in the native state and then carry out the same digest after modifying the protein, for example by methylation. Furthermore, since lysine (K) and arginine (R) are two of the most common amino acids in proteins, there is a relatively large number of doublets and triplets (e.g. RR, KKR, KRK). Trypsin cleaves randomly at such sites generating peptide fragments with **ragged termini**. More confidence can be placed in database hits that are compatible with such peptides because this is effectively the same as searching with a larger number of peptides.

Database search tools

A number of algorithms have been developed for sequence database searching using MS data. Among these, the most commonly used is **SEQUEST**, which works by searching for all peptides in the specified database(s) with the same mass as a given peptide ion. Then, a virtual digest is performed on the matched protein and a theoretical mass spectrum generated. The data from the theoretical mass spectrum are then compared to the experimental data and the best matches are scored. In a different approach, peptide mass data are used to generate a collection of possible sequences, and this profile is used as a query in a modified BLAST or FASTA search. A program called **Lutkefisk** has been developed for this purpose. Many software resources for the analysis of MS data are available over the Internet and some are listed in *Table 1*.

Limitations of MS analysis

Although a powerful technique for protein annotation and sequencing, there are some limitations to MS, which need to be taken into account when interpreting and analyzing experimental data. One of the most important factors to take into account is that MS data may not match any database entry due to the presence of an unknown **post-translational modification**. Where such modifications are known, exact mass differences between unmodified and modified amino acids can be predicted. Indeed several algorithms, including SEQUEST, have built-in parameters for detecting such modifications. However, the presence of a modified residue should always be confirmed experimentally. Another potential problem is the occurrence of nonspecific proteolysis. This depends on the purity of the cleavage agent used. Many algorithms will carry out peptide mass searches without a specified cleavage agent to take nonspecific proteolysis into consideration. A common problem is that protein spots isolated from 2D-PAGE

Table 1. Internet resources for MS-based protein identification

Resource	URL	Features and comments
CBRG, ETH-Zurich	cbrg.inf.ethz.ch/Masssearchtml	Peptide mass search
European Molecular Biology Laboratory (EMBL), Heidelberg	www.mann.embl-heidelbergde/Services/ PeptideSearch/PeptideSearchIntro.html	Peptide-mass and fragment ion search
ExPASy	www.expasy.ch/tools/#proteome	Peptide-mass and fragment ion search
Mascot	www.matrix-science.com/cgi/index.pl?page/ home.html	Peptide-mass and fragment ion search
Rockefeller University New York	prowl.rockefeller.edu/	Peptide-mass and fragment ion search
SEQNET, Daresbury, UK	www.seqnet.dl.ac.uk/Bioinformatics/ welapp/mowse	Peptide-mass and fragment ion search
University of California	prospector.ucsf.edu	Peptide mass (MS-Fit) and fragment ion (MS-Tag) search
	donatello.ucsf.edu	As above, on-line access
University of Washington	Thompson.mbt.washington.edu/ sequest/	Instruction on how to get the SEQUEST fragment-ion search program

Modified from Pennington S.R. and Dunn M.J., *Proteomics: From Protein Sequence to Function*, © BIOS Scientific Publishers Limited, 2001, p. 99.

gels often contain a mixture of proteins, and these contaminants may be difficult to identify. Finally, imperfect matches may result because the actual protein does not exist in the database, but a close homolog from the same species or a different species, which may have a related sequence, does exist. This is often the case if a protein contains a single nucleotide polymorphism leading to two or more variants. Programs such as MS-BLAST and CID entify use sequence candidates from tandem mass spectrometry data as the input for a homology search (Topic E1).

L1 MODELING AND RECONSTRUCTING MOLECULAR PATHWAYS

Key Notes

Higher-order systems	Although genes and proteins can be studied individually, more insight into their functions can be gained by studying higher-order systems, that is, molecular pathways and networks, cells, tissues, organs and whole organisms. This allows their physical and functional interactions to be determined in the widest possible context.
Representation of pathways and networks	Molecular pathways and networks can be represented by graphs, with molecules at the nodes and relationships shown by links. In metabolic pathways, nodes represent substrates or intermediates and links represent their catalytic interconversion by enzymes. In signaling and regulatory pathways, nodes represent proteins and links indicate the transfer of information. Graphs of molecular pathways are generally directional and can show positive and negative interactions.
Reconstruction of molecular pathways	Pathways and networks can be mapped directly by substrate feeding experiments and *in vitro* enzyme assays. More recently, a number of indirect but high-throughput methods have been developed thanks to the advent of functional genomics. These methods include pathway reconstruction from expression data, protein interaction data and comprehensive mutagenesis programs.
Modeling molecular pathways	Mathematical models of biochemical reactions are often based on differential equations that predict the change in concentration of particular molecules over time. Simultaneous differential equations can be used to model entire pathways and several software applications are available for this task, including GEPASI and BioQuest. There are limitations to the use of simultaneous differential equations and these have been addressed through the development of stochastic models based on the Gillespie algorithm, which is incorporated into programs such as StochSim.
Molecular pathway resources	There are many resources for viewing molecular pathways on the Internet. One of the most comprehensive for metabolic pathways is KEGG and this also shows a selected range of regulatory pathways. An important feature of such resources is that the contents of the maps are integrated with other databases by way of hyperlinks.
Related topics	Phylogenetics, cladistics and ontology (G1) Protein interaction bioinformatics (L2)
	Gene expression data: tools and resources (J2) Higher-order models (L3)

Higher-order systems

Genes and proteins can be studied as individual entities. This is exemplified by the annotated sequence databases discussed in Topic C2. These databases store both sequence data and text annotations for individual genes. The annotations provide additional (generally functional) information about genes and their products. Although such resources are invaluable, they represent the sequence–function relationships of single molecules, that is, individual components of a biological system. To gain a full insight into the function of proteins, it is necessary to assimilate these data into **higher-order systems**, which include pathways, networks, cells, tissues, organs and whole organisms.

In this topic, we consider the role of bioinformatics in the study of molecular pathways and networks. A **molecular pathway** is a group of proteins that function in series to achieve a common objective. For example, a **metabolic pathway** processes a substrate through a series of intermediates to generate a product, a **signaling pathway** propagates a signal from a receptor to its target(s) and a **regulatory pathway** represents a hierarchy of control. A **molecular network**, in this context, is a higher-level system that may encompass several or many pathways and their interactions. Bioinformatics is important for the representation, reconstruction and modeling of pathways and networks, and in the creation of database resources.

Representation of pathways and networks

The simplest way to represent a molecular pathway is to use a graph. As discussed in Topic G1, a graph shows the relationship between entities (represented by nodes) using connecting lines as links. The entities in this case are molecules, and the links show some kind of physical or functional relationship.

In the case of a metabolic pathway, the nodes show metabolic substrates, intermediates and products while the links show how these molecules are interconverted. The links might be labeled with the enzymes responsible for catalyzing each reaction step. In the case of a signaling cascade, the nodes represent components of the cascade, for example the signal itself, the receptor, cytoplasmic signal transduction proteins and downstream targets. The links indicate the signal transduction route and generally infer that adjacent components in the pathway physically interact. This would not necessarily be the case for enzymes catalyzing successive steps in a metabolic pathway since the metabolic intermediates could diffuse or be transported between different enzymes. In the case of a regulatory pathway, the nodes indicate the regulatory molecules, for example transcription factors, and the links represent some hierarchy of control.

Graphs of molecular pathways and networks range from the simple to the very complex. They may incorporate multiple branches, convoluted junctions and circuits. Furthermore, they usually show directionality in the relationship between nodes, for example representing the direction of metabolic flux or signal transduction. Graphs of regulatory pathways and networks must also show whether interactions are positive or negative. Positive interactions (stimulation or activation) are usually shown as arrows and negative relationships (inhibition or inactivation) are usually shown as T-bars.

Reconstruction of molecular pathways

Molecular pathways can be reconstructed or mapped using a variety of direct and indirect methods. Perhaps the most direct method is to carry out **feeding experiments** with labeled metabolic intermediates. Incorporation of the label into the final product of the pathway is then proof of the intermediacy of the introduced compound. The isolation of enzymes and their use in *in vitro* **enzyme assays** is another approach that may yield useful information. More recently, functional genomics and bioinformatics have provided a number of tools that facilitate pathway reconstruction. **Gene expression data** can be analyzed and used to build models of pathways and networks based on **common expression profiles** (see Topic J2 for more details) and **protein interaction data** may provide supporting evidence (Topic L2). The analysis of **mutants** is also useful. If several genes have very similar mutant phenotypes it suggests the proteins may be involved in a common pathway. The order of gene action in the pathway can be established by genetic analysis.

Modeling molecular pathways

Given the appropriate information, it is possible to derive mathematical models of biochemical reactions that predict the levels of reactants and products over time. A simple model might involve the formation of a product P from a substrate S. The rate of change in the level of product would be dependent on the concentration of S, that is, [S], and could be described by the following equation where V_{max} is the maximum velocity and K_M is the Michaelis constant:

$$dP/dt = V_{max}[S]/([S]+K_M)$$

Expanding on this idea, an entire metabolic pathway can be modeled using a series of differential equations, and the model can be modified to cope with complicating factors such as branch points, shuttling of intermediates between cell compartments and spontaneous (i.e. non-catalyzed) reactions.

There are many software applications that can be used to model metabolic pathways in this manner. For example, **GEPASI** (which can be downloaded from the following World Wide Web (WWW) site: http://www.gepasi.org) and **BioQUEST Metabolic Pathways/Enzyme Kinetics Simulation** (which can be downloaded from the following WWW site: http://omega.cc.umb.edu/ ~bwhite/ek.html). Both these programs simulate the kinetics of biochemical reactions and allow models to be built of quite complex systems, including those regulated by feedback from molecules in the pathway. GEPASI runs only on Windows 9x/NT computers. BioQUEST is a set of building blocks that run on the numerical simulation program Extend.

Models based on simultaneous differential equations are not ideal because they make implicit assumptions about the state of the system, for example that the reaction is at steady state and that all enzymes are equally accessible. An alternative approach uses the **Gillespie algorithm**, which is not based on concentrations but on particles that react randomly according to a reaction probability calculated from known rate constants. This generates a **stochastic simulation** of the chemical reaction kinetics. The Gilespie algorithm has been incorporated into programs such as **StochSim**, a biochemical simulator in which molecules are represented as individual software objects. This program was developed to provide a more realistic model of bacterial chemotaxis and is able to represent multiple post-translational modifications and conformational states of protein molecules. It can be downloaded from the following WWW site: http://www.zoo.cam.ac.uk/comp-cell/StochSim.html.

Molecular pathway resources

A number of excellent molecular pathway databases can be found on the Internet. The great advantage of these over traditional references sources, such as biochemistry text books, is that the components of the pathways are cross-referenced by way of hyperlinks to data stored in other databases such as GenBank, SWISS-PROT and organism-specific databases (Topics C2 and C3).

One of the most extensive resources is the **Kyoto Encyclopedia of Genes and**

Genomes (KEGG). The top page is located at the following URL: http://www.genome.ad.jp/kegg/. By clicking on *Open KEGG* the user is taken to the contents page, at the top of which is *Pathway Information* with links to *Metabolic Pathways* and *Regulatory Pathways*. Clicking on *Metabolic Pathways* brings up an extensive list of pathways under various subheadings. Clicking on a subheading gives a broad outline metabolic map with links to more specific pathways. If the user clicks on one of these (either from the map or from the text link on the previous page), a detailed graph is shown of the appropriate pathway, with substrates, intermediates and products as nodes, links showing the direction of catalysis, and boxes showing the enzymes represented by their EC numbers. The user can click on any enzyme, and this initiates a LinkDB search for information in associated KEGG databases (see Topic D1). For example, a user interested in taurine metabolism can scroll down the contents page until the link *Taurine and hypotaurine metabolism* is found under *Metabolism of Other Amino Acids*. Clicking on this link brings up the graph that is reproduced in *Fig. 1*. The circles are links to adjacent pathways, that is, those either supplying substrates to or using products from taurine and hypotaurine metabolism. Clicking on the title of the graph initiates a LinkDB search and lists information on components of the pathway in other databases.

The *Regulatory Pathways* is not such a comprehensive and exhaustive resource as the *Metabolic Pathways*, but nevertheless provides a wealth of information. The type of information provided varies from cases to case. Clicking on *Basal transcription factors*, for example, brings up pictures of macromolecular complexes and links to sequence database entries for each component. Clicking on *Cell cycle* reveals a more traditional graph with proteins at the nodes and links showing regulatory interactions. A very useful feature of this system is that the LinkDB report for individual components also points to related records in other databases, which may have further pathway information. For human records, for example, there are links to OMIM (Topic C4) and for *Drosophila* records, there are links to Flybase and GADFly (Topic C3).

The KEGG top page also contains a *Link to Pathway and Other Databases*. This reveals a list of resources including eight other pathway databases, which the reader can explore for his or her self. One of these alternatives is the metabolism section of the ExPASy site (Topic A3), which can be reached at the following URL: www.expasy.ch/cgi-bin/search-biochem-index. This contains digitized versions of the Boehringer Mannheim 'Biochemical Pathways' and 'Cellular and

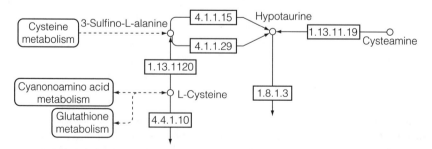

Fig. 1. Taurine and hypotaurine metabolism entry in the KEGG metabolic pathways database. Links are shown by rectangles with rounded corners (to adjacent pathways) and sharp corners (to enzyme database entries). The title is also a link to a list of database entries relevant to the entire pathway.

Molecular Processes' wall charts, each of which can be zoomed as appropriate to look at specific pathways. Alternatively, the user can enter a search term and be taken directly to the appropriate region. Using the same example as above, if taurine is used as a search term, a list of matches is returned (*Fig. 2*) linking to different parts of the map. The enzymes on the map are hyperlinks to the matching entries in the ENZYME database.

 Biochemical Pathways

Taurine found

The decimal classification refers to the "Enzyme Nomenclature, Recommendations (1984)" (Academic Press, Orlando etc. 1984) and its supplements. If you click on one of these EC numbers, you will get the corresponding entry of the ENZYME Data Bank.

Besides, for each item of the index, the following list contains the numbers(s) of all squares of the map "Biochemical Pathways" in which this item appears.

HYPOTAURINE
Available map(s): *I4*

HYPOTAURINE DEHYDROGENASE ENZYMEL *1.8.1.3*
Available map(s): *I4*

TAURINE
Available map(s): *I4, K10*

Spektrum Courtesy of Spektrum Akademischer verlag, Gmbh
© 1993 Boehringer Mannheim GmbH - Biochemica

Fig. 2. Result of searching the ExPASy metabolism site with 'taurine'. The numbers (I4, K10) refer to grid references on the digitized Boehringer Mannheim 'Biochemical Pathways' wall chart.

L2 PROTEIN INTERACTION INFORMATICS

Key Notes

Interactions and pathways
Proteins that physically interact with each other may be involved in the same molecular pathway or network, or may form part of a multisubunit complex. Using this principle, pathways can be reconstructed based on evidence of protein interactions. However, information from other sources – for example gene expression patterns and mutant phenotypes – may also be useful.

Handling Y2H data
Yeast two-hybrid (Y2H) screens produce large amounts of protein interaction data, but there is a relatively high level of spurious results (false positives and false negatives). This problem can be addressed by scoring interactions for reliability, based either on the repeatability of interactions over multiple experiments, or by the number of times a given bait will trap independent clones representing the same prey. Even so, similar large-scale screens tend to identify different (although overlapping) sets of interactions.

Protein interaction databases
Several databases have been set up to store the interaction data arising from large-scale Y2H screens. However, much more information on protein interactions is available in the scientific literature and a current challenge in bioinformatics is the assimilation of these interaction data from diverse sources.

The interactome
The interactome is the sum of all protein interactions in the cell. The simplest way to represent protein interactions is a graph with proteins as nodes and interactions as links. However, when large numbers of proteins are considered, the graphs become too complex. They can be simplified by clustering functionally similar proteins, resulting in a functional interaction map that links fundamental cellular processes.

Related topics: Protein interaction data (B4)   Modeling and reconstructing molecular pathways (L1)

Interactions and pathways
As discussed in Topic L1, the function of individual proteins can only be fully elucidated by studying them in the context of pathways and other higher-order systems. Pathways can be reconstructed from different types of functional data, including the study of mutants and gene expression patterns (see Topic J2 for methods). Protein interaction data, however, provide not only a functional basis but also a physical basis for pathway reconstruction. Protein interaction data may also demonstrate the existence of multisubunit complexes. In this topic, we consider the role of bioinformatics in the handling and analysis of protein inter-

action data and show how this helps in the definition of molecular pathways, networks and complexes.

Handling Y2H data

Most of the techniques used to study protein interactions (Topic B4) are low-throughput methods, that is, they are most suitable for studying the interactions of a single protein, or interactions among a small group of proteins such as a multisubunit complex. For these methods, the challenge of bioinformatics is to find ways of assimilating the data from diverse individual experiments into a single resource (see below). Conversely, library based methods – particularly the yeast two-hybrid (Y2H) system – are geared for high-throughput screening and thus generate large data sets. The challenge for bioinformatics in this context is to mine the data and extract meaningful results.

The greatest problem with Y2H screens is the relatively high proportion of spurious results. These comprise **false positives**, where the reporter gene is activated in the absence of any specific interaction between the bait and prey, and **false negatives**, when the reporter gene is not activated even if the bait and prey do normally interact, as shown by other experimental techniques. False positives occur for many reasons, for example if the prey is **sticky** (interacts nonspecifically with many proteins) or capable of **autoactivation** (activation of transcription without interacting with the bait). False negatives often occur if the fusion protein does not fold in the same manner as the native protein.

The analysis of Y2H data thus requires each result to be assessed for reliability. Before considering how confidence is assessed, an appreciation of the Y2H assay format is required. Essentially, large-scale Y2H screens come in two types. In a **matrix assay**, each bait construct and each potential prey construct is individually produced, for example using the polymerase chain reaction, and haploid yeast strains are arranged as a matrix on a series of microtiter plates allowing systematic and exhaustive crossings to be carried out, therefore testing each possible interaction. Conversely, in a **random library assay**, the bait constructs are produced individually but the prey are represented by a library of random DNA clones. Libraries of high complexity are required to cover an entire genome. In a matrix assay, reliability is generally judged by the repeatability of a particular interaction. That is, the crossings are carried out a number of times, and only those interactions that are highly repeatable are taken to be genuine results. In a random library assay, each prey is generally represented by a number of overlapping clones. Reliability can be judged by the number of independent clones, representing the same prey, which are trapped by the bait. A further advantage of this approach is that the domain of the prey protein responsible for the interaction can be narrowed down (*Fig. 1*). Despite these precautions, it is apparent that similar Y2H screens tend to identify different sets of interacting proteins. For example, between 1999 and 2001, two research groups carried out large-scale interaction screens covering the entire yeast genome (see Further Reading). In total, over 1500 high-confidence interactions were catalogued, but less than 10% of these were identified in both screens.

Protein interaction databases

Protein interaction data are very useful for functional annotation and the reconstruction of molecular pathways and networks. One role of bioinformatics is to provide **protein interaction databases** that allow interaction data to be stored, queried, assessed for confidence and used for pathway reconstruction. Several databases have been set up to store the results from individual large-scale Y2H screens, while others are more general in their scope (*Table 1*). Despite the

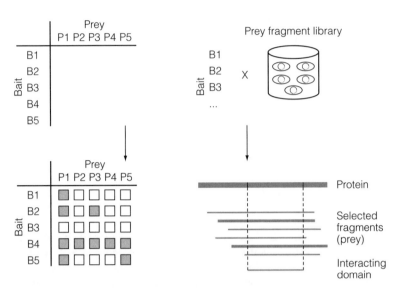

Fig. 1. *Principle of the matrix and random library formats for Y2H screens. In the matrix format, different bait/prey combinations are tested systematically. In the library format, coverage of the genome is dependent on library complexity but the domains of interacting proteins can be identified by the analysis of overlapping prey clones. Reproduced with permission from Legrain P. et al., Protein–protein interaction maps: a lead towards cellular function,* Trends Genet **17:** *348, 2001.*

Table 1. *A selection of protein interaction databases available over the Internet*

Database	URL	Comments
BIND	http://www.bind.ca	Biomolecular Interaction Network Database. Lists protein interactions with a variety of molecules (proteins, nucleic acids, small ligands) and includes resources for protein complexes and molecular pathways
DIP	http://dip.doe-mbi.ucla.edu	Database of Interacting Proteins. Comprehensive resource for protein-protein interactions
Hybrigenics	www.hybrigenics.com	Lists proteins interactions in the bacterium *Helicobacter pylori*. Includes graphical interface for the visualization of interactions
Pronet	http://pronet.doubletwist.com	Lists human protein interactions
	http://portal.curagen.com/ extpc/com.curagen.portal.servlet	Two yeast protein interaction databases based on individual genome-wide Y2H screens
	http://genome.c.kanazawa-u.ac.jp/Y2H	

amount of data generated in Y2H experiments, there is a much larger volume of protein interaction data already in the scientific literature, representing the results of thousands of individual experiments on specific proteins. A number of software tools have been developed that scan literature databases such as MEDLINE and attempt to extract useful information on protein interactions through the analysis of titles, keywords and abstracts.

The interactome   The ultimate challenge for bioinformatics in the field of protein interaction technology is the reconstruction of the **interactome**, defined as the sum of all protein interactions in the cell. The presentation of these data in an accessible and user-friendly format is very difficult. The simplest way to represent protein interactions is a mathematical graph called an **interaction map**, with proteins at the nodes and interactions represented by links. For a small number of proteins, interaction maps are very useful. However, larger numbers of proteins yield graphs of incredible complexity. For example, the graph shown in *Fig. 2* represents the interactions of just 1500 yeast proteins. The yeast genome actually comprises over 6000 genes, so at least four times the number of proteins shown would have to be included to cover the entire interactome.

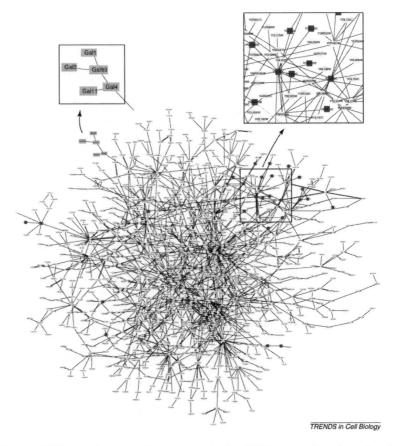

*Fig. 2. A highly complex interaction map representing 1500 yeast proteins. Reproduced with permission from Tucker C.L. et al., Towards an understanding of complex protein networks, Trends Cell Biol **11:** 102, 2001.*

Simplification can be achieved by clustering functionally similar proteins, that is, allocating proteins to functional categories such as DNA replication, transcription, membrane transport etc. The end result is a **functional interaction map** in which basic cellular functions are linked together in a network. Such a map for the yeast interactome is shown in *Fig. 3*. Inspection of the map

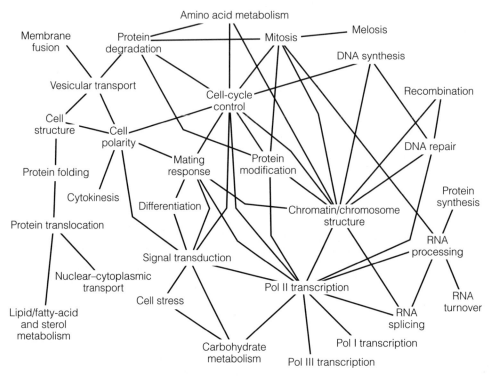

*Fig. 3. A functional interaction map based on Fig. 2. The proteins have been clustered into functional centers allowing all basic cellular functions to be linked in a network. Reproduced with permission from Tucker C.L. et al. Towards an understanding of complex protein networks, Trends Cell Biol **11**: 103, 2001.*

shows that certain functional classes of protein tend to interact with each other. For example, proteins involved in DNA replication interact with each other, but also with proteins involved in DNA repair, since the two processes are related. Conversely, proteins involved in the control of mitosis do not generally interact with those involved in vesicular transport, since the two processes are discrete. By building such an interaction map, the validity of newly discovered interactions can be tested. In particular, implausible interactions revealed by Y2H screens can be tested more rigorously using other experimental techniques.

L3 HIGHER-ORDER MODELS

Key Notes

The cell

The cell can be regarded as a compartmentalized set of molecular pathways and networks distributed in space and restricted by membranes. Any model of a cell must incorporate these features. A useful modeling resource is Virtual Cell, in which the cell is defined as a collection of structures, molecules, reactions and fluxes. The user can define biological or mathematical models for cell function.

Tissues and organs

Tissues and organs comprise organized populations of interdependent cells. Modeling depends on an accurate description of the geometry of the tissue and must include any time-dependent processes. For example, modeling the heart requires a description of its anatomy and the way in which action potentials are propagated. The model must take into account the fact that cardiac muscle is an anisotropic system.

Organisms

In order to model an entire organism, it is necessary to have a sound understanding of the principles underlying development. For most multicellular organisms there is too little information and the developmental program too complex for this to be achieved. The nematode *C. elegans*, however, has a number of features that make it an ideal system upon which to base a developmental model. It is a simple organism (it has about 1000 somatic cells) whose somatic cell lineage is invariant, making perturbations in development very easy to identify. Models of *C. elegans* development have been generated based on the concept of three spaces: genomic space, cellular space and developmental space.

Related topics

Modeling and reconstructing molecular pathways (L1)

Protein interaction informatics (L2)

The cell

Molecular pathways and networks, as discussed in Topic L1, represent only the first level of higher-order system in which genes and proteins are considered in terms of functional and interacting groups rather than as single entities. The next level is the **cell**, which in the simplest sense can be regarded as a collection of interrelated and compartmentalized molecular pathways, including those responsible for exchanging information with the environment, separated in space by membranes. Defining the cell, however, is not simply a case of working out and organizing all the molecular pathways encoded by the organism's genome. The cell is a dynamic system in which some pathways are active and some are silent depending on environmental conditions. In multicellular organisms, cells differentiate and specialize so only a subset of the potential pathways encoded by the genome is used in any one cell type. Therefore, systematic attempts to catalogue metabolic, signaling and regulatory pathways (Topic L1) or networks of protein interactions (Topic L2) are not sufficient for modeling the cell.

More sophisticated approaches involve the modeling of physiological processes in the context of three-dimensional cell structure. The **Virtual Cell** is an excellent modeling utility, which allows the cell to be defined as a collection of structures, molecules, reactions and fluxes. The user can define cellular compartments and membranes, fill particular compartments with specific molecules, specify channels and pumps that allow the movement of molecules between compartments, and establish biochemical reactions. The reactions are defined in terms of kinetics, stoichiometry and location. The Virtual Cell should appeal to both biologists and mathematicians since it is possible for the user to create biological models from which the software will generate the mathematical code needed to run simulations, or to use the Virtual Cell's own language to define a mathematical model for cell function, whose results can be analyzed as images. The Virtual Cell can be accessed at the following URL, and includes a useful tutorial for the inexperienced user: http://www.nrcam.uchc.edu.

Tissues and organs

Tissues and organs represent higher-level systems than the cell. A tissue is an organized collection of cells of one or more types. An organ is a structure that has adapted to fulfil a specialized function in the body. Any modeling approach must take into account the fact that cells within tissues and organs interact with each other in a variety of ways (e.g. by the release of short-range diffusible molecules or by contact-dependent mechanisms) and may be interdependent. Formally, the function of one cell may be dependent on the function of another, so the cells are no longer considered as individual functional entities but as components of the system. Another feature of tissues and organs is that any perturbation of their normal biochemistry or physiology is generally met by an attempt to restore homeostasis. To a certain extent this is also true of cells, but cells can also react to changes in the environment by undergoing differentiation.

Applied examples of tissue and organ modeling can be found on the Internet. For example, a model for the simulation of ventricular propagation can be found at the following URL: http://www.cbiol.leeds.ac.uk/arun/reconstruct.html. The aim of this research is to construct a model of the heart that is useful for virtual experiments on heart function and dysfunction. The model consists of a description of heart geometry and ordinary differential equations to describe action potentials. The model is complex because cardiac tissue is an anisotropic excitable system, that is, waves of action potentials have different magnitudes in different directions.

Organisms

The modeling of entire (multicellular) organisms involves a similar set of problems to the modeling of tissues and organs, that is, an appreciation of the communication, cooperation and interdependency of different cell types. A unique problem applicable at the organism level, however, is the modeling of development. The **developmental program**, as it is sometimes called, is written into the genome and contains the information required to convert a single cell – the fertilized egg – into a complex functioning organism in which there are many different cell types organized in a specific manner to form defined structures (tissues and organs). In most organisms, there is too little information available about the underlying molecular processes to build comprehensive developmental models[1] but the nematode worm *Caenorhabditis elegans* is excep-

---

[1] This situation is likely to change as more functional data is derived from the genome projects.

tional. This organism has several properties that make it uniquely suitable as a basis for modeling development:

- *C. elegans* is a widely-studied model organism favored by developmental biologists.
- The *C. elegans* genome has been sequenced; indeed, it was the first genome of a multicellular organism to be sequenced.
- The somatic cell lineage is invariant, that is, development consists of a series of stereotypical cell divisions, so all cells can be traced back through the developmental hierarchy to the egg.
- As a consequence of this, the embryonic body plan is invariant: each hermaphrodite individual has exactly 959 somatic cells arranged in exactly the same way, and each male has exactly 1051 somatic cells arranged in exactly the same way.
- This makes it relatively easy to identify mutations that affect development, since such mutations result in specific changes to the number and/or organization of cells.
- It is also relatively easy to study the physiology of this organism. Therefore, a complete wiring diagram of the *C. elegans* nervous system is available.

An example of a *C. elegans* developmental model can be found at the following URL: http://www.bionet.nsc.ru/bgrs/thesis/23/. This model is based on the concept of three levels or spaces: **genomic space**, which relates gene expression to protein function (metabolism, signal transduction etc.), **cellular space**, which describes the cell lineage, and **developmental space**, which describes the spatial arrangement of cells and their behavior (*Fig. 1*).

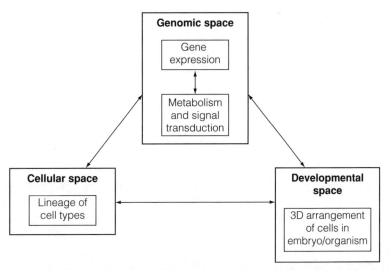

Fig. 1. Relationships among the 'three spaces' which can be used in models of developing multicellular organisms.

M1 CONVENTIONS FOR REPRESENTING MOLECULES

Key Notes

Cheminformatics

Cheminformatics, a combination of chemistry and information technology, is required for the processing and analysis of chemical data. Cheminformatics is relevant to biologists because chemistry data are important in many areas of molecular biology, for example in the study of protein interactions and metabolism.

Molecular formulae

Molecules can be represented by simple formulae, which give the number and type of atoms. However, this does not show how they are connected. Structural formulae provide some information about the arrangement of atoms in a molecule and thus allow isomers to be distinguished.

Structural diagrams

Molecules can be represented using simple graphs, which show atoms as nodes and bonds as links. For organic molecules, further simplification is achieved by assuming that carbon atoms make up the molecular backbone and that the valency of four is satisfied by hydrogen atoms unless otherwise shown. Such diagrams present all molecules as planar shapes and do not indicate the spatial distribution of atoms in three dimensions.

Chirality

If four different groups are coordinated around a central carbon atom, the molecule is described as chiral. Chiral molecules exist in two conformations, enantiomers, which are mirror-images of each other. Although enantiomers have the same chemical properties, many enzymes and other proteins show chiral selectivity, which is important in drug development and related fields. Molecules may contain any number of chiral centers and a series of forms, called diastereoisomers, may exist. These may have different chemical properties because of the way different groups interact within the molecule.

DL and *RS* conventions

The absolute configuration of groups around a chiral carbon atom can be described using a number of conventions. In the DL system, molecules are named D or L according to whether the coordinated groups are arranged in a similar fashion to those in D-glyceraldehyde or L-alanine. In the *RS* system, molecules are named *R* (*rectus*) or *S* (*sinister*) according to the size of the chemical groups surrounding the carbon atom.

Related topics

Protein interaction bioinformatics (L2)

Cheminformatics resources (M2)

Bioinformatics and drug discovery (N1)

Pharmainformatics resources (N2)

Cheminformatics   Bioinformatics traditionally deals with the structure and function of large molecules, that is, nucleic acids and proteins. However, small molecules also play an
important role in the cell, interacting with macromolecules and acting as
substrates, ligands, cofactors and regulators. **Cheminformatics** is the combination of chemistry and information technology that deals with the analysis of
such molecules. The subject appears in this book because chemistry data are
important in many areas of biology, for example in the kinetics of biochemical
reactions and in the interactions between proteins and small molecules. In this
topic, we briefly discuss conventions for the unambiguous representation of
small molecules. In Topic M2, we explore selected cheminformatics resources
available for the biologist. Such resources include data formats, software applications for modeling and viewing molecules, and databases of chemical structures. Algorithms for docking small molecules into the binding sites of proteins
are discussed in Topic N2.

Molecular   The term 'small molecule' can mean anything from simple inorganic
formulae   compounds (e.g. water, sodium chloride) to complex organic compounds, that
is, molecules that are based on a carbon skeleton. Irrespective of the type of
molecule, however, there must be a standard nomenclature that allows such
molecules to be described unambiguously. There are several different ways of
representing molecules – names, simple formulae, structural formulae,
structural diagrams and text strings. All these systems have advantages and
disadvantages.

Molecules can be represented using **simple formulae** that list the number and
type of atoms. This is sufficient for very simple compounds such as water (H_2O),
carbon dioxide (CO_2) and sodium chloride (NaCl), because the atoms can only
be arranged in one manner so the structure of the molecule is implicit. However,
ambiguity arises where the same atoms can be arranged in several different
ways. For example, both ethanol and dimethyl ether have the simple formula
C_2H_6O, but they have different structures. They are known as structural
isomers. Such ambiguities can be resolved through the use of **structural
formulae** that show not only the number and type of atoms but also how they
are connected. Using this type of representation, ethanol is C_2H_5OH and
dimethyl ether is CH_3OCH_3.

Structural   From a structural formula, it is possible to draw a basic **structural diagram** of
diagrams   any molecule. The most straightforward way to achieve this is to represent
molecules as graphs, with atoms identified by their chemical symbols at the
nodes and chemical bonds shown as links. Further simplification is achieved in
organic molecules by representing carbon atoms as simple nodes (no chemical
symbol) and assuming that the valency of four is satisfied with hydrogen atoms
unless otherwise shown. Some examples of this convention are shown in *Table 1*.

Although such conventions are useful for describing molecular structures,
they assume that all molecules are planar (i.e. will lie flat on a surface). In actual
fact, the four bonds surrounding a saturated carbon atom are distributed tetrahedrally in space. A more detailed diagram of the ethane molecule is shown in
Fig. 1. In panels *a* and *b*, note that the carbon atoms, shown as nodes, are in the
plane of the paper, as are the hydrogen atoms linked to the carbons by ordinary
lines. The hydrogen atoms shown in bold are above the plane of the paper, and
the bold lines represent bonds coming out of the page. Conversely, the dotted
lines represent bonds going into the page and joining to hydrogen atoms that

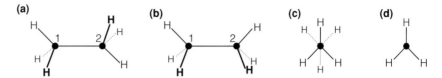

Fig. 1. Structural representations of ethane that show the tetrahedral distribution of coordinated groups about the saturated carbon atoms. Panels (a) and (b) show two extreme conformations. The energetically favorable conformation (a), which predominates in nature, has hydrogen atoms on opposite sides of the C–C bond as far as possible from each other (in the staggered configuration). The less favorable conformation (b) has the atoms in the eclipsed configuration. Panels (c) and (d) show the same conformations viewed from the end of the molecule.

Table 1. Structural formulae and full and simplified structural diagrams for some common organic compounds.

Name	Formula	Full structure	Simplified structure
Methane	CH_4	H | H—C—H | H	
Ethane	C_2H_6	H H | | H—C—C—H | | H H	————
Ethene (ethylene)	C_2H_4	H 〉C=C〈 H H	=
Ethanol	C_2H_5OH	H H | | H—C—C—O | | H H	/OH
Cyclohexane	C_6H_{12}	H H C—C H H H—C-H H-C—H H H C—C H H	⬡
Phenol	C_6H_5OH	H H C=C H—C C—OH C—C H H	—OH
Ethanal (acetaldehyde)	CH_3CHO	H H | | H—C—C=O | H	/O

are below the plane of the paper. Panels *a* and *b* show alternative conformations of ethane caused by rotation of the C–C bond. Any sample of ethane will contain a dynamic mixture of these extreme conformations and all possible intermediates, with the energetically favorable conformation shown in panel *a* predominating. Panels *c* and *d* show the same conformations as *a* and *b*, but viewed from the end of the molecule. The significance of this in the context of representing molecular structures will become clear below.

Chirality

Many naturally occurring organic substances are **optically active**, that is, in solution they rotate the plane of polarized light. Historically, such substances were classified as **dextrorotatory** (+) if the plane of light was rotated rightwards or **levorotatory** (−) if it was rotated leftwards. There are remnants of this system in such familiar biochemical terms as L-alanine and D-glucose. The actual direction of light rotation has no structural significance, but the D/L system has been retained as a convention for describing the 'handedness' of molecules with an **asymmetric carbon atom**. This phenomenon is known as **chirality**.

A molecule is chiral if the four groups coordinated to a saturated carbon atom are all different. Note that carbon atoms are not the only atoms that act as chiral

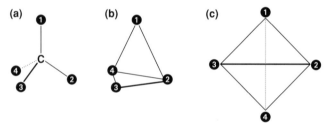

Fig. 2. Representation of a tetrahedrally coordinated saturated carbon atom in an organic molecule. (a) The carbon atom is at the center of a tetrahedron with four coordinated groups (❶❷❸❹); (b) simplified representation with the central carbon removed; (c) representation of the tetrahedron as a flat image.

(a) $CH_2OHCHOHCHO$

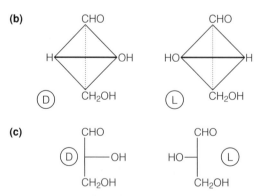

Fig. 3. (a) The structural formula of glyceraldehyde gives no indication of its chirality; (b) if the molecule is represented as a tetrahedron, the D and L enantiomers can be distinguished; (c) these can be shown as two-dimensional graphs using the Fischer convention.

centers. Nitrogen, phosphorus and sulfur can also act as chiral centers in organic molecules, and metal ions can act as chiral centers when they chelate other molecules in solution. In such situations the groups can be arranged in two alternative configurations that are mirror images of each other, and cannot be superimposed. These alternative, mirror image molecules are known as **enantiomers**. For the purpose of representing such molecules, it is convenient to view a carbon atom as the center of a tetrahedron, as shown in *Fig. 2a*. This can be simplified by removing the carbon atom (*Fig. 2b*) and laying the structure flat (*Fig. 2c*). A real case is shown in *Fig. 3*. The structural formula of glyceraldehyde is shown in *Fig. 3a*. This allows the structure of the molecule to be drawn but gives no indication of its handedness. However, if the second carbon atom is viewed as the center of a tetrahedron, the molecule has two enantiomers, D-glyceraldehyde and L-glyceraldehyde (*Fig. 3b*). These structures can be distinguished in a two-dimensional graph if the asymmetric bonds are shown pointing in different directions, and this is known as the **Fischer convention** (*Fig. 3c*). As usual, the presence of the fourth bond, to a single hydrogen atom, is assumed but not explicitly shown.

A molecule may contain two or more chiral centers, a phenomenon termed **diastereoisomerism**. In this case a series of different forms may exist with alternative conformations. Familiar examples of diastereoisomeric molecules are L-threonine and D-glucose (*Fig. 4*).

Of what practical significance are enantiomers and diastereoisomers? Enantiomers are mirror images of each other but differ only in their optical properties. Diastereoisomers, however, also differ in their chemical and physical properties because the chemical groups distributed around the chiral centers may interact in different ways. Importantly, many enzymes and other proteins demonstrate chiral preference with the effect that biological systems are biased towards particular conformations. For example, chemically synthesized amino acids occur as a **racemate**, that is, an equal mixture of D and L forms.

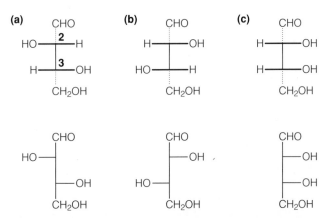

Fig. 4. (a) D-erythrose; (b) L-erythrose; (c) D-threose. These are all four-carbon sugars with the formula $CH_2OH(CHOH)_2CHO$. The second and third carbons are chiral centers. The top panel shows full structural layouts and the bottom panel shows the same structures using the Fisher convention. D-erythrose and L-erythrose are enantiomers because the molecular configurations are mirror images of each other. D-threose is a diastereoisomer of both D-erythrose and L-erythrose because the configuration of one chiral carbon has changed but the other remains the same. D-erythrose and D-threose are so-called because they have the same absolute configuration as D-glyceraldehyde (Fig. 3c).

Conversely, amino acids synthesized in biological systems are almost exclusively L-enantiomers, while sugars are usually D-enantiomers. These issues have to be taken into account in drug development. The activity of drugs, including antibiotics, is highly dependent on their chirality because they bind to chiral receptors.

DL and *RS* conventions

Although optical rotation can be used to discriminate between enantiomers, the direction of rotation is not related to the absolute configuration of groups around a chiral center and is therefore of little use for systematic nomenclature. Several conventions have arisen for the precise definition of molecular conformations, and two of these – the DL and *RS* conventions – are widely used. The **DL convention** is based on the structures of D-glyceraldehyde and L-alanine. Structures are described as D if the arrangement of groups around a chiral carbon atom is similar to that in D-glyceraldehyde and as L if the arrangement is similar to that in L-alanine. This is irrespective of the optical properties of the molecule. The *RS* convention, also known as the **CIP (Cahn–Inglod–Prelog) convention** after its inventors, is a systematic nomenclature based on the sizes of the chemical groups surrounding the central carbon atom. *R* and *S* stand for the Latin words *rectus* and *sinister*. Both these systems are used in databases of small molecules so that the absolute configuration is explicitly stated (Topic M2). The complete rules can be found in any standard textbook of organic chemistry (see Further Reading).

M2 CHEMINFORMATICS RESOURCES

Key Notes

SMILES

SMILES is a system for representing chemical formulae as strings, based on a valence model in which all valencies are considered to be satisfied by hydrogen atoms unless otherwise shown. The system has conventions for representing different bond types, cyclic molecules, branches, *cis/trans* isomers and chirality.

RasMol and Chime

There are several specialized data formats for chemical structures based on the principle of a molecular formula and associated table of connections. Viewing utilities such as RasMol and Chime can interpret these file formats and display interactive molecular structures in a variety of user-defined schemes and colors.

Chemical structure databases

Structural information about different molecules can be obtained from a number of comprehensive WWW resources, including Chemical Abstracts On-Line, Chemfinder and MedChem. Each of these resources provides a chemical database that can be searched using a variety of query formats, for example systematic name, nonsystematic name, formula, molecular weight or CAS registry number. Search results provide physical, chemical and biomedical information with links to other databases and resources. MedChem also provides the SMILES string.

QSAR

A QSAR is a statistical method used to determine how the structural features of a molecule are related to biological activity. The QSAR approach is particularly useful for categorizing the activities of related molecules with multiple functional groups. Each molecule is broken down into a series of descriptors (molecular properties) and the QSAR determines which descriptors are most likely to promote biological activity. This gives rise to a set of rules that can be used to evaluate the potential activity of new molecules.

Related topics

Conventions for representing molecules (M1)

Bioinformatics and drug discovery (N1)

Pharmainformatics resources (N2)

SMILES

For the storage and exchange of data on chemical structures, a simple yet comprehensive nomenclature is desirable. Several workable systems have been developed but here we will concentrate on one of the most popular, **SMILES (Simplified Molecular Input Line Entry Specification)**, in which chemical structures are represented by text strings. For the interested reader, a complete tutorial can be found at the following URL: http://www.daylight.com/

dayhtml/smiles/smiles-intro.html#TOC. The discussion below provides a brief overview of the SMILES system.

In SMILES, molecules are represented according to a **valence model**, in which the valency of each atom is considered to be satisfied by hydrogens unless otherwise shown. In this context, the rules are similar to those in the simplified structural diagrams discussed in Topic M1. For example, the SMILES format for methane (CH_4) is simply C, since in the absence of any further information all four bonds are assumed to be satisfied with hydrogen atoms. Similarly, the SMILES format for water (H_2O) is O and that for ammonia (NH_3) is N. Ethanol (CH_3CH_2OH) is represented by the string CCO.

Further symbols and conventions are required to represent different bond types. A double bond is represented by two parallel lines, hence molecular oxygen is O=O and carbon dioxide is O=C=O. Hydrogen cyanide is C#N, where # represents a triple bond. Cyclic molecules are indicated by numbering the atom at which the ring is closed, so cyclohexane is C1CCCCC1. Special rules apply to $sp2$ hybrid carbon atoms which feature in aromatic compounds. These can be written as lower case letters, so benzene can be shown as c1ccccc1 and pyridine as n1ccccc1. However, the same molecules can also be represented by showing a mixture of single and double bonds, that is, benzene can alternatively be shown as C1=CC=CC=C1 and pyridine as N1=CC=CC=C1. Branches in molecules are indicated by parentheses, thus $(CH_3)_2CHCO_2H$ (isobutyric acid) is written as CC(C)C(=O)O.

Discrimination between *cis* and *trans* isomers is achieved by using the / and \ symbols to show that atoms are arranged either on opposite sides or on the same side of a double bond. For example, *trans*-dibromoethene would be shown as Br/C=C/Br (or Br\C=C\Br) because the bromine atoms are on opposite sides of the double bond. Conversely, *cis*-dibromoethene would be shown as Br/C=C\Br (or Br\C=C/Br) because the bromine atoms are on the same side of the double bond.

A special convention is used to distinguish enantiomers (Topic M1). As an example consider the molecule fluoroalanine, which has an amino group, a methyl group, a fluorine atom and a carbonyl group arranged around the chiral carbon. The general SMILES format for this molecule (without specifying chirality) would be NC(C)(F)C(=O)O. However, since the first carbon atom in the string is a chiral center, there are two enantiomers. To distinguish these molecules, the chiral carbon atom is enclosed in square brackets. Then the first coordinating group in the string (in this case the amino group, N) is used as a reference point (*Fig. 1*). Viewing the molecule from this reference point, the remaining groups are arranged either clockwise or counterclockwise to satisfy the order in which they appear in the string (methyl, fluorine, carbonyl). These alternatives are represented by the symbols @@ (clockwise) and @ (counter-clockwise), that is, N[C@@](C)(F)C(=O)O and N[C@](C)(F)C(=O)O. A special case arises where one of the coordinating groups is a hydrogen, since hydrogens are not usually explicitly shown in SMILES nomenclature. Where a hydrogen must be specified to define the chirality of the molecule, it is included within the square brackets with the chiral carbon. Therefore L-alanine is written thus: N[C@@H](C)C(=O)O. D-alanine is written thus: N[C@H](C)C(=O)O.

RasMol and Chime

An alternative to the representation of molecules as text strings is to expand upon the concept of the molecular formula and to associate this with a table of connections. In a simple case, the formula C_2H_6O could represent ethanol

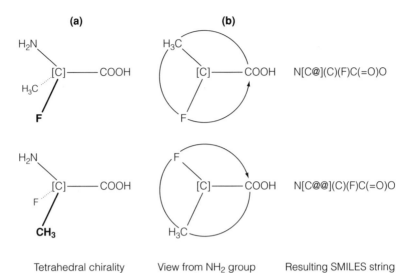

(a)   (b)

Tetrahedral chirality   View from NH_2 group   Resulting SMILES string

Fig. 1. Method used in SMILES to specify the chirality of an organic molecule, in this case fluoroalanine. (a) Three-dimensional structural diagrams showing the D and L enantiomers of this molecule. (b) Structures viewed from the amino group, showing the three remaining groups arranged either clockwise or anticlockwise to satisfy the order of groups in the SMILE string.

unambiguously if followed by a table or matrix summarizing the connections within the molecule. There are several specialized data formats for use in chemistry and biochemistry that follow this principle. There are also a number of file-conversion utilities that allow structures encoded in these formats to be translated into standard structural diagrams, viewed and manipulated. Perhaps the most widely used of these utilities is **RasMol** (rastering molecules), a stand-alone program made freely available over the Internet by its author. Since the source code of this program is in the public domain, it has been modified for use on a wide range of computer systems. A similar program is **Chime**(chemical Multiple Internet Mail Extensions; MIME), which can be downloaded for free at the MDL (Molecular Design Ltd.) WWW site (http://www.mdli.com). Part of the Chime program is based on RasMol, so the user interface is similar. However, Chime runs as a browser plug-in application rather than a stand-alone program. Both RasMol and Chime accept molecular coordinate files in a number of formats, including the format used by the Protein DataBank (Topic C2). They display interactive molecules on-screen in user-defined colors and various schemes (including framework, space-filling, macromolecular ribbon and dot-surface representations). The interactive user interface allows the molecules to be rotated and zoomed, and the user can add text annotations.

Chemical structure databases

There are several excellent WWW sites providing searchable databases of chemical structures. A very comprehensive resource is the online version of **Chemical Abstracts** (http://www.cas.org/casdb.html), which also includes many biological and medical entries. Another useful facility is **Chemfinder** (http://chemfinder.cambridgesoft.com), a database of chemical structures and associated information that can be searched by name, formula, molecular weight or CAS registry number. A typical Chemfinder search result is shown in *Fig. 2*. The search term used was *quercetin*, one of many nonsystematic names for a

Quercetin [117-39-5]

Synonyms: 3,3',4',5,7-pentahydroxyflavone; 2-(3,4-dihydroxyphenyl)-3,5,7-trihydroxy-4H-1-benzopyran-4-one;
3,5,7,3',4'-pentahydroxyflavone; 3',4',5,7-tetrahydroxyflavon-3-ol; cyanidelonon 1522; C.I. natural yellow 10; C.I. natural yellow 10 & 13; C.I.
natural red 1; C.I. 75670; meletin; quercetol; quertine; sophoretin; t-gelb bzw. grun 1; xanthaurine;

$$C_{15}H_{10}O_7$$
302.2398

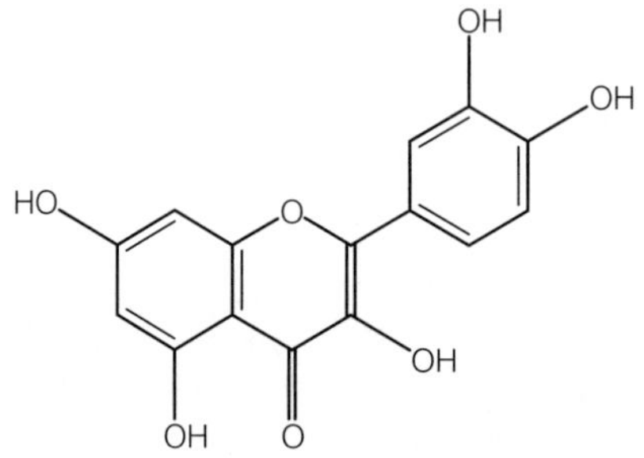

There follow the accession number, certain physical properties, the Chemical Abstracts number (117-39-5) and links to relevant topics as below.

Biochemistry
Ligand Chemical Database for Enzyme Reactions
 Information about this particular compound
 Information about this particular compound

Chemical Online Order
Available Chemicals Exchange
  Information about this particular compound

Health
9 links to toxicology, carcinogenecity and mutagenicity

Medications
Phys. Nutrition Encyclopedia
  Information about this particular compound

Physical Properties
63 structural descriptors for NTP compounds
Proton NMR Spectral Molecular Formula Index
  Information about this particular compound

Regulations
3 links to regulations that might apply to quercetin

Fig. 2. Chemfinder search results, using the term quercetin as a search query.

natural yellow pigment that can be extracted from plants. Information retrieved in this search included the fact that the three-dimensional structure is known, and that the molecule is of interest in medicine, nutrition and plant molecular biology. Each of these annotations includes links to accessory sources of information. One detail not provided by Chemfinder is the SMILES string for quercetin. This can be found at an alternative resource, **MedChem** (http://clogp.pomona.edu/medchem/chem/master/search.html). These search facilities are very user friendly and tolerate typographical variations, errors and synonyms.

QSAR

There are many situations in which it would be useful to be able to predict the activity of a particular molecule based on its structure. In drug development, for example, lead compounds are optimized by decorating the molecular skeleton with different functional groups and testing each derivative for its biological activity (Topic N1). If there are several open positions on the lead molecule that can be substituted, the total number of molecules that need to be tested in a comprehensive screen would be very large. The synthesis and screening of all these molecules would be time-consuming and laborious, especially since most would have no useful activity. Under these circumstances, a QSAR can be used to select those molecules most likely to have a useful activity, and therefore guide chemical synthesis. A **QSAR** is a **quantitative structure–activity relationship**, a mathematical relationship used to determine how the structural features of a molecule are related to biological activity. Essentially, molecules are treated as groups of molecular properties (**descriptors**), which are arranged in a table. The QSAR mines these data and attempts to find consistent relationships between particular descriptors and biological activities, thus identifying a set of rules that can be used to score new molecules for potential activity.

A QSAR is usually expressed in the form of the linear equation shown below, where P_1–P_N are parameters (molecular properties) established for each molecule in the series and C_1–C_N are coefficients calculated by fitting variations in the parameters to their biological activities:

$$\text{Biological activity} = \text{Constant} + \sum_{i=1}^{i=N} C_i \cdot P_i$$

In this context, 'biological activity' might include such factors as therapeutic index, toxicity or carcinogenicity. A variety of statistical and computational methods is used for optimizing the parameters and these include adaptive techniques such as genetic algorithms. A very readable introduction to the QSAR method showing a worked example for the molecule capsaicin is available at the following URL: http://www.netsci.org/Science/Compchem/feature19.html.

N1 BIOINFORMATICS AND DRUG DISCOVERY

Key Notes

Drugs	Drugs interact with targets, usually proteins, in the body and through such interactions cause physiological responses. The pharmaceutical industry aims to discover drugs with specific beneficial effects to treat human diseases.
Principles of drug development	Drug development begins with the identification of a suitable target, which must contribute significantly to a human disease. Ideally, altering the activity of this target should have a beneficial effect thus showing its potential for therapeutic intervention. The next stage of the process is lead discovery, where compounds showing some of the desired activity of an ideal drug are sought. Optimization of lead compounds results in drug candidates that may be registered and submitted for clinical trials, which establish their safety and metabolic behavior in human subjects.
Bioinformatics in drug development	Genomics, proteomics, combinatorial chemistry and HTS have all contributed to a massive increase in the amount of data generated by the pharmaceutical industry. The role of bioinformatics is to store, track and provide tools for the analysis of these data. Specific applications include the modeling of protein interactions with small molecules allowing rational drug design, the association of genotype and drug response patterns (pharmacogenomics), the design and assessment of chemical diversity in combinatorial libraries, and the processing and storage of data from high-throughput screens of lead compounds.
Related topics	Microarray data: analysis methods (J1) Protein interaction informatics (L2)
	Analyzing data from 2D-PAGE gels (K1) Cheminformatics resources (M2)
	Modeling and reconstructing molecular pathways (L1) Pharmainformatics resources (N2)

Drugs

A **drug** is a molecule that interacts with a **target** in the body (a biological molecule, usually a protein) thus triggering a physiological effect. Drugs can be beneficial or harmful depending on that effect. Drugs for the treatment of diseases interact with targets that contribute to the disease and produce positive effects. The disease target may be endogenous (a protein synthesized by the individual to whom the drug is administered) or, in the case of infectious diseases, may be produced by a pathogenic organism. Drugs act by either stimulating or blocking the activity of the target protein.

Principles of drug development

The development of a new drug is a complex, lengthy and expensive process. Current estimates place the total development costs of a new drug (including the writing off of false starts, clinical trials and tests required by regulatory authorities) somewhere in the region of 800 million dollars. The viability of the exercise requires a company to be able to launch one or two new drugs per year whereas the actual figures are typically 0.4–0.8 new drugs per year per company. This is one reason for the consolidation (through mergers and takeovers) of the large pharmaceutical companies.

Drug development begins with the identification of a potentially suitable disease target, a process called **target identification**. In the past, target identification was based largely on medical need. Nowadays, with many useful therapies already on the market or in development, target identification depends not only on medical need but also on factors such as the success of existing therapies, the activity of competing drug companies and commercial opportunities. The next stage of the process is **target validation**, which involves extensive testing of the target molecule's therapeutic potential. Validation may include the creation of animal disease models (Topic N2) and the analysis of gene and protein expression data under various conditions (see Sections J and K). A valid target must contribute to a human disease and have a high **therapeutic index**. This means that, should an appropriate drug be found, a significant therapeutic gain must be predicted through the use of such as drug.

Once a target has been validated, the search begins for drugs that interact with that target. This process is called **lead discovery**, and involves the search for **lead compounds**, that is, substances with some but not all of the desired biological activity of the ideal drug. **Lead optimization** involves the modification of lead compounds to produce derivatives, **candidate drugs (CDs)**, with better therapeutic profiles. For example, increased specificity for the target would be desirable because this would result in reduced side-effects. The CDs are then assessed for quality, taking into account factors such as ease of synthesis and formulation (conversion into a capsule, pill or cream, etc.).

Development candidates passing this test may then be registered as an **investigational new drug** and submitted for **clinical trials**. This is the lengthiest and most expensive part of the drug development process, which is why most projects are abandoned before this stage. Clinical trials are designed to determine safety and tolerance levels in humans, and to discover how the drug is metabolized. Trials are divided into several stages:

- Preclinical phase: animal studies.
- Phase I: normal (healthy) human volunteers.
- Phase II: evaluation of safety and efficacy in patients, and selection of dose regimen.
- Phase III: large patient number study with placebo or comparator; at this stage regulatory approval is sought and a commercial launch decision taken.
- Phase IV: long-term monitoring for adverse reactions reported by pharmacists and doctors.

Bioinformatics in drug development

Over the last 10 years, drug development has benefited greatly from advances in biology and chemistry, predominantly in three areas: genomics/proteomics, combinatorial chemistry and high-throughout screening. Due to these advances, the scale and complexity of target identification/validation and lead discovery/optimization experiments has increased dramatically, resulting in huge amounts of data that need to be stored and analyzed. The term **pharmainformatics** is often used to describe the mix of biology, chemistry, mathematics and information technology required for data handling in the pharmaceutical industry. The scope of pharmainformatics is outlined in *Table 1*.

Genomics and proteomics have revolutionized the way target molecules are identified and validated. Traditionally, drug targets have been characterized on an individual basis and lead compounds sought with specific clinical effects. With the advent of genomics, particularly the availability of the entire human genome sequence and its annotations, thousands of potential new targets can now be identified by sequence, structure and function. Bioinformatics is important not only because of its role in the analysis of sequences and structures, but also in the development of algorithms for the modeling of target protein interactions with drug molecules. This allows **rational drug design**, in which protein structural data (Section I) is used to predict the type of ligands that will interact with a given target, and therefore form the basis of lead discovery. We return to this subject in Topic N2.

Another new development based on genomics is the emerging scientific field of **pharmacogenomics**. This is the study of the association between genomic data and drug response patterns. Some individuals with a disease may benefit from a drug while others receive little or no benefit or may experience drug-

Table 1. Areas of biology and chemistry where informatics plays a vital role in the drug discovery pipeline

Application	Role of bioinformatics
Biology	
Genomics/proteomics (human genome project)	
Characterization of human genes and proteins	Target identification/validation in the human genome
	Cataloging SNPs and association with drug response patterns (pharmacogenomics)
Genomics/proteomics (human pathogen genome projects)	
Characterization of the genes and proteins of organisms that are pathogenic to humans	Target identification/validation in pathogens
Functional genomics (protein structures)	
Analysis of protein structures (humans and their pathogens)	Prediction of drug/target interactions
	Rational drug design
Functional genomics (expression profiling)	
Determining gene expression patterns in disease and health	Gene classification based on drug responses
	Pathway reconstruction
Functional genomics (genome-wide mutagenesis)	
Determining the mutant phenotypes for all genes in the genome	Databases of animal models
	Target identification/validation
Functional genomics (protein interactions)	
Determining interactions among all proteins	Characterization of protein interactions
	Reconstruction of pathways
	Prediction of binding sites
Chemistry	
HTS	
Highly parallel assay formats for lead identification	Storing, tracking and analyzing data
Combinatorial chemistry	
Synthesis of large numbers of chemical compounds	Cataloging chemical libraries
	Assessing library quality/diversity
	Predicting drug/target interactions

related side-effects. Such individual variations may be due to underlying variation in the structure of the target molecule, or differences in drug metabolism,
both of which depend on genotype. Variations in drug response patterns are
responsible, in part, for the high failure rates of new drug candidates at the clinical trials stage. **Genomic medicine** is the concept of treating patients with
drugs tailored to their genotype, which can be rapidly screened by searching for
individual changes in the nucleotide sequence, known as **single nucleotide
polymorphisms** (**SNPs**, pronounced 'snips').

The way in which lead compounds are identified has also changed significantly over the last 10 years. Traditionally, there are four sources of lead
compounds:

- Lead compounds may be discovered through **chance observations** (the most
  famous example of this was the discovery of penicillin, which has given rise
  to many of today's antibiotics, by Alexander Fleming).
- They may also be identified on the basis of **side-effects** caused by drugs
  developed for alternative treatments (e.g. the tranquilizer chlorpromazine
  was discovered following its experimental use as an antihistamine).
- Lead compounds may be derived from **traditional medicines**, for example
  quinine from cinchona bark.
- Lead compounds may be **natural substrates or ligands** (e.g. epinephrine
  (adrenaline) which has given rise to analogs such as salbutamol, which is
  used to treat asthma).

More recently, these traditional approaches have been joined by more systematic methods for lead identification. These methods are based on **high-
throughput screening (HTS)** in which lead discovery is accelerated through the
use of highly parallel assay formats, such as 96-well plates. In turn, this requires
the assembly of large chemical libraries for testing. This has been made possible
by **combinatorial chemistry** approaches, in which large numbers of different
compounds can be made by pooling and dividing materials between reaction
steps. Bioinformatics is important in all these developments: for the storage,
tracking and analysis of HTS data, and for the design and assessment of
chemical libraries. These subjects are also addressed in Topic N2.

The impact of genomics, proteomics, combinatorial chemistry and HTS on
drug development is difficult to judge, simply because the time from target
identification to the marketing of a successful new drug is so long. However,
these advances increase the number of compounds that can be screened by
several orders of magnitude and should also make it easier to choose between a
potentially successful drug and a likely nonstarter. More than 99% of all drug
development projects are abandoned before the new drug is marketed and only
a third of marketed drugs actually recuperate the costs of research and development. Although these high-throughout methods will increase the research and
development budgets of the pharmaceutical companies significantly, there
should also be a good return in the form of marketable novel drugs.

N2 PHARMAINFORMATICS RESOURCES

Key Notes

Pharmainformatics	Pharmainformatics is the combination of biology, chemistry, mathematics and information technology that is essential for efficient data management, processing and analysis in the pharmaceutical industry.
Chemical diversity	Diverse chemical libraries are required for efficient lead discovery if little is known about the binding properties of the drug target. Conversely, focused libraries are required if the structure of the target is known, since this defines a particular set of ligands. Chemical diversity can be defined by comparing molecules on the basis of descriptors (functional groups) and how these fill chemical space. A number of software tools are available for the design and assessment of diverse or focused chemical libraries, and virtual screening against drug targets.
Computational screening	Software applications such as DOCK and Autodock match potential ligands to binding sites by calculating steric constraints and bond energies. These can be used to search chemical databases and find potential drug leads. Some applications consider the ligand and binding site as inflexible structures, rather like pieces of a jigsaw, while others can incorporate flexibility into the molecules by calculating allowable and compatible bond torsions.
Functional genomics	The large-scale functional annotation of genes is known as functional genomics and incorporates areas such as homology searching, structural analysis, expression analysis, large-scale mutagenesis and the analysis of protein interactions. All of these areas are important in drug development. Genome-scale mutagenesis is a rich source of animal disease models for target identification and validation, and large mutant collections in simple organisms can be used for the rapid high-throughput screening of potential lead compounds.
Pharmacogenomics	Pharmacogenomics is the study of how variation in the human population correlates with drug response patterns. The analysis of genomic data and its comparison with drug response data allows patients to be clustered into drug response groups, so that appropriate drugs and dose regimens can be administered. Variation is catalogued by analyzing data on mutation (particularly SNPs) and gene expression profiles.

Related topics:	Sequence similarity searches (E1) Conventions for representing
	 Microarray data: analysis methods molecules (M1)
	    (J1) Cheminformatics resources (M2)
	 Protein interaction informatics (L2) Bioinformatics in drug development
	                                               (N1)

Pharmainformatics The combination of biology, chemistry, mathematics and information technology required for data processing and analysis in the pharmaceutical industry is sometimes termed **pharmainformatics**. The need for pharmainformatics in drug development was discussed briefly in Topic N1, reflecting the impact of genomics, proteomics, combinatorial chemistry and high-throughput screening (HTS) on drug discovery. In this topic, we consider specific applications of pharmainformatics in more detail and discuss available database and software resources.

Chemical diversity HTS in drug discovery depends on the availability of diverse chemical libraries, such as those generated by combinatorial chemistry, since these maximize the chances of finding molecules that interact with a particular target protein. It is difficult to quantify **chemical diversity** but attempts have been made based on the concept of 'chemical space'. Essentially, **chemical space** encompasses molecules with all possible chemical properties in all possible molecular positions. A diverse library would have broad coverage of chemical space, leaving no gaps and having no clusters of similar molecules. In practice, library diversity is quantified using measures that compare the properties of different molecules based on **descriptors** such as atomic position, charge and potential to form different types of chemical bond. Two molecules can be compared using the **Tanimoto coefficient (T_C)**, which evaluates the similarity of fragments of each molecule. The coefficient is calculated as shown below where a is the number of fragment-based descriptors in compound A, b is the number of fragment-based descriptors in compound B and c is the number of shared fragment-based descriptors:

$$T_C = c/(a+b-c)$$

Thus, for identical molecules, $T_C = 1$, while for molecules with no descriptors in common, $T_C = 0$. In a chemical library of ideal diversity, most pair-wise comparisons would generate a Tanimoto coefficient near to zero.

Diverse libraries are very useful for lead discovery when little is known about the binding specificity of the target protein. However, it is more often the case that some form of sequence or structural information is available for the target, and this can be used in the design of **focused libraries** that concentrate on one region of chemical space. For example, if the sequence of a particular target protein is known, then database homology searching (Topic E3) will often find a related protein whose structure has been solved and whose interactions with small molecules have been characterized. In such cases, it is possible to design a chemical library based on a particular **molecular scaffold**, which preserves a framework of sites present in a known ligand, but which can be **decorated** with diverse functional groups. Some of these groups may have previously been shown to be important for drug binding (such sites are known as **pharmacophores**).

Tools and resources are available for the design of combinatorial libraries and the assessment of chemical diversity. For example, a program called **Selector**, available from Tripos, allows the user to design very diverse libraries or libraries focused on a particular molecular skeleton. **Chem-X**, developed by the Oxford Molecular Group, allows the chemical diversity in a collection of compounds to be measured and identifies all the pharmacophores. Another Tripos program, **CombiLibMaker**, allows a virtual combinatorial library to be generated and used to screen a virtual target. Similar resources are available from MDL Information Systems.

Computational screening

Laboratory-based screening experiments are very expensive, so it is beneficial to generate as much information about potential drug/target interactions as possible before such an experiment is undertaken. The **computational screening** of chemical databases, using a target molecule of known structure, is one way in which such information can be obtained. The structure of the target molecule may be known from X-ray crystallography or nuclear magnetic resonance spectroscopy data (Topic B2). Alternatively, the solved structure of a close homolog may be used, or the structure may be predicted using a threading algorithm (Topic I7). If the structure of a target protein is known, algorithms can be used to identify potential interacting ligands based on goodness of fit, therefore allowing **rational drug design**. Several well-known drugs have been developed in this manner, including Relenza and Captopril.

Many **docking algorithms** have been developed which attempt to fit small molecules into binding sites using information on steric constraints and bond energies. A list of free software resources and associated URLs is provided in *Table 1*. One of the most established docking algorithms is **Autodock**. Initially the software comprised three modules, Autotors, Autogrid and Autodoc. Autotors dealt with ligand co-ordinates and the definition of bonds that were axes of free rotation, Autogrid constructed a three-dimensional grid of interaction energies, and Autodock performed the docking simulation itself. Another widely used program is **DOCK**, in which the arrangement of atoms at the

Table 1. Chemical docking software available over the Internet

URL	R/F	Description	Availability
http://www.scripps.edu/pub/olson-web/doc/autodock/index.html	F	Autodock, discussed in main text	Download for UNIX/LINUX
http://swift.embl-heidelberg.de/ligin/	R	LIGIN, a robust ligand–protein interaction predictor limited to small ligands	Download for UNIX or as a part of the WHATIF package
http://www.bmm.icnet.uk/docking/	R	FTDock and associated programs RPScore and MultiDock. Can deal with protein-protein interactions. Relies on a Fourier transform library.	Download for UNIX/LINUX
http://reco3.musc.edu/gramm/	R	GRAMM (Global Range Molecular Matching), an empirical method based on tables of inter-bond angles. GRAMM has the merit of coping with low-quality structures.	Download for UNIX or Windows.
http://cartan.gmd.de/flex-bin/FlexX	F	FlexX, which calculates energetically favorable molecular complexes consisting of the ligand bound to the active site of the protein, and ranks the output.	Apply on-line for FlexX workspace on the server.

R/F means rigid or flexible, and indicates whether the program regards the ligand as a rigid or flexible molecule.

binding site is converted into a set of spheres called **site points**. The distances between the spheres are used to calculate the exact dimensions of the binding site, and this is compared to a database of chemical compounds. Matches between the binding site and a potential ligand are given a confidence score, and ligands are then ranked according to their total scores. This modeling approach has the disadvantage that the binding site and every potential ligand are considered to be stiff and inflexible. Other algorithms (see *Table 1*) can incorporate flexibility into the structures. A more recent development called **CombiDOCK** considers each potential ligand as a scaffold decorated with functional groups. Only spheres on the scaffold are initially used in the docking prediction and then individual functional groups are tested using a variety of bond torsions. Finally, the structure is **bumped** (checked to make sure none of the positions predicted for individual functional groups overlap) before a final score is presented.

Chemical databases can be screened not only with a binding site (searching for complementary molecular interactions) but also with another ligand (searching for identical molecular interactions). Several available algorithms can compare two-dimensional or three-dimensional structures and build a profile of similar molecules. This approach is important, for example, if a ligand has been shown to interact with a protein but has negative side-effects, or if a structurally distinct ligand is required in a drug development project to avoid intellectual property issues. In each case, a molecule of similar shape with similar chemical properties is required, but with a different structure.

Functional genomics

Functional genomics incorporates any branch of biology dealing with the high-throughput functional annotation of genes, that is, the determination of gene functions on a genomic scale (*Table 2*). Functional genomics is set to have a massive impact on the pharmaceutical industry because it provides a fast track to target identification and validation. Thus far, there are no examples of drugs

Table 2. *Approaches used in functional genomics, all of which are relevant to the pharmaceutical industry. Bioinformatics plays an important role in each of these fields, either through the development of algorithms and statistical methods for data analysis, or through the development of resources for data management*

Approach	Functional annotation method	Related topics
Homology searching	Comparison to related sequences with known function	Sections E and F
Protein structure determination (structural genomics)	Comparison to molecules with related structure and known function	Sections F and I
Comparative genomics	Functional annotation by domain conservation, conserved phylogeny or conserved genomic organization	Section G
Expression analysis	Similar expression profiles indicate conserved function	B3, J1, J2, K1, K2
Mutagenesis	Function based on mutant phenotype, e.g. knockout mice	This topic
Protein interaction screening	Function based on presence in multi-subunit complex or on interaction with proteins of known function	B4, L2
Small molecule informatics	Interaction with small molecules	M1, M2

that have been developed solely on the basis of functional genomics because the field is new, but the impact has already been felt in the early stages of drug development and it is likely that the output of the industry, in terms of newly registered drugs, will increase dramatically in the next 5–10 years.

Bioinformatics and functional genomics go hand in hand because the very nature of functional genomics means that large amounts of data are generated. Some aspects of functional genomics are covered elsewhere in this book, specifically the prediction of protein function by sequence comparison (Sections E and F) and structural analysis (Section I), the analysis of gene and protein expression data (Sections J and K) and the analysis of protein interaction data (Section L). All of these play a major role in the pharmaceutical industry. One aspect of functional genomics that is of particular relevance to drug development, and that has not been discussed elsewhere, is **genome-wide mutagenesis**, the deliberate creation of mutants for each and every gene in the genome.

In the pharmaceutical industry, the mouse is the most important animal because its physiology and biochemistry are so similar to those of humans. Drugs can be tested on healthy mice and on those in which the disease in question has been artificially induced (such an animal is known as a **disease model**). In many cases, disease models are produced by genetic manipulation, generating either a **transgenic mouse** (a mouse with additional genetic material) or a **knockout mouse** (a mouse in which a particular gene has been mutated, disrupted or removed by homologous recombination). In the past, disease models have been created on an individual basis by targeting particular genes. Over 15 000 different transgenic and knockout mouse strains have been generated in individual projects, providing the material for many potential disease models. These strains are listed in a number of dedicated databases, the most comprehensive of which are **T-BASE** (http://tbase.jax.org) and the **BioMedNet knockout mouse database** (http://biomednet.com/db/mkmd). Disease models can also be generated by **random mutagenesis** rather than the creation of specific strains. Large-scale mutagenesis of the mouse has been carried out using a chemical mutagen called **ENU (ethylnitrosourea)**. Over 40 000 mutagenized mouse strains have been generated and scored for phenotypes relevant to drug development, including allergy, immunology, clinical chemistry, nociception (response to pain), dysmorphology (abnormal development) and behavior. Another method for the generation of random mutants is **insertional mutagenesis**, where a DNA construct is integrated randomly into the genome and may disrupt a gene. Currently, two organizations, the **German Gene Trap Consortium (GGTC)** and the US biotechnology company **Lexicon Genetics** are carrying out projects to generate genome wide mutant libraries in which every possible mouse gene is interrupted. Mutagenesis is carried out in **embryonic stem cells**, which are derived from the early mouse embryo and have the potential to generate all cell types. Banks of mutant ES cells can be stored indefinitely at low temperatures. At the appropriate time, the cells can be thawed and injected into normal mouse embryos where they colonize and contribute to the germ line, allowing the production of transgenic mice in the subsequent generation. So far more than 20 000 mutant cell lines have been produced in the two projects. Further information can be found on the Internet, at the resources listed in *Table 3*.

Bioinformatics has many roles to play in the management of genome-wide mutagenesis projects. The cataloging of mutant strains and the tracking of their behavior under many experimental conditions is one example. The creation of

Table 3. Internet resources for large-scale mutagenesis projects in the mouse. Bioinformatics is important for database management and gene identification by sequence analysis

Project	URL
Transgenics and knockouts	
Jackson laboratory web site (2500 strains of knockout mutant mice listed)	http://jaxmice.jax.org/index.shtml
TBASE, database of transgenics and knockouts	http://tbase.jax.org
BioMedNet Mouse Knockout Database	http://biomednet.com/db/mkmd
ENU mutagenesis	
German Human Genome Project ENU mutagenesis	http://www.gsf.de/ieg/groups/enu-mouse.html
UK/French consortium ENU mutagenesis project	http://www.mgu.har.mrc.ac.uk/mutabase
Insertional mutagenesis	
Lexicon Genetics 'OmniBank' ES cell library	http://www.lexgen.com/omnibank/omnibank.htm
German Gene Trap Consortium ES cell library	http://tikus.gsf.de

databases allowing mouse strains to be searched by gene, phenotype and clinical relevance is another. Random mutagenesis projects involve a further issue, which is the identification of the interrupted genes. The GGTC and Lexicon Genetics maintain databases of **flanking sequence tags**, short DNA sequences derived from the flanks of their integrated DNA constructs, which can be searched against GenBank to help identify the gene that is interrupted.

It should be noted that, although the mouse is the most relevant model organism to humans, genes and proteins do tend to be highly conserved across the animal kingdom. The functional similarity between gene products in humans and other animals is so conserved that some biotechnology and pharmaceutical companies are using a factory-style screening approach in which thousands of mutant strains of a simple organism, for example the nematode worm *Caenorhabditis elegans*, are tested with a panel of potential drugs to identify interactions that rescue or otherwise influence the mutant phenotype. The advantage of this approach is the highly parallel assay format and the fact that these simple organisms have complete genome sequences with highly informative annotations. The mutant gene in *C. elegans* can thus be identified rapidly and used to identify homologous genes in the human genome that may act as novel drug targets.

Pharmacogenomics As discussed in Topic N1, **pharmacogenomics** is the study of association between genome data and drug response patterns. Pharmacogenomics is an important new development because it offers the promise of **personalized medicine**, that is, drugs targeted to an individual's genome. This should help to avoid the clinical problem of **adverse drug reaction**, a severe negative reaction to drug administration that is thought to kill as many as 100 000 people a year. There is currently no easy method for predicting an individual's response to drug treatment, so pharmaceutical companies are restricted to the development of drugs that are suitable for all patients, which severely limits their scope. In the future, pharmacogenomics may allow the development of more specialized and targeted drugs.

Bioinformatics plays a vital role in pharmacogenomics because it allows data on genetic variation to be compared meaningfully with data on drug response patterns. Correlations, generated by clustering similar data, can divide patients into discrete **response groups**. For example, there are four recognized response groups to the drug nortriptyline: poor metabolizers, intermediate metabolizers, extensive metabolizers and ultra-rapid metabolizers. These phenotypic groups reflect allelic variants of the *CYP2D6* locus, which encodes an enzyme of the cytochrome P450 oxidase family. Patients in each group are given different doses of nortriptyline commensurate with their metabolic characteristics.

The analysis of variation among the human population is facilitated by databases that catalog mutations and expression patterns. Expression databases are discussed in Topic J2, and include the National Center for Biotechnology Information (NCBI) Gene Expression Omnibus (GEO). The NCBI holds a second database, **dbSNP**, which lists human **single nucleotide polymorphisms (SNPs).** These are individual nucleotide changes which, when they occur in protein-coding genes, may be associated with many different phenotypes including variations in the absorbance or clearance of drugs. The rationale behind the collection of SNPs is the creation of a large data set that will allow profiles of SNPs to be built up and matched against particular drug response patterns. In the future, this will allow individual patients to be 'typed' for their likely response to drug administration, and SNP data is likely to be incorporated into the early stages of drug development (target validation) as well as forming an important part of the clinical trials stage. The NCBI SNP database can be found at the following URL: http://www.ncbi.nlm.nih.gov/SNP/.

01 RUNNING COMPUTER SOFTWARE

Key Notes

Software frameworks for bioinformatics	A computer can be thought of as comprising hardware (physical devices such as processors and disk drives) and software (programs). The system software is the computer's master program, that is, the operating system, whereas application software carries out specific tasks for the user. Application software may be installed and executed locally, remotely over a network or over the Internet.
Programs and programming languages	A computer may be programmed in machine code, which is the language used by its processor, or in one of a range of higher-level languages that must be assembled or compiled into machine code prior to execution. Executable files (in Widows) or executable images (in UNIX systems) are precompiled and stored in the computer's memory. Remote software may have to be compiled as it is running.
Scripts and scripting languages	Scripts are program files that are executed by another program rather than by the computer's processor. These are versatile for both stand-alone and remote applications and are written in scripting languages such as PERL and JavaScript.
Popular languages in bioinformatics	Bioinformatics is typically a cross-platform discipline in which computers are used to integrate and manage data from a variety of sources. Popular programming, scripting and markup languages in bioinformatics reflect this versatility, and include the programming language Java, the scripting languages PERL and JavaScript, and the markup languages HTML and XML.
Running programs over the Internet	Programs can be executed over the Internet either client-side or server side. Client-side applications are typically embedded in HTML and are supplied for example as JavaScript or Java applets. Server side applications are run at the server using a CGI and can be in a variety of programming or scripting languages. Client-side software may have to be downloaded, but runs independently of server performance. The performance of server side software depends on server load.
Related topics	Bioinformatics and the Internet (A3) Software downloading and installation (O3) Computer operating systems (O2)

Software frameworks for bioinformatics

Software is a collective term for the various different **programs** that can run on computers. Software is distinguished from **hardware**, which refers to physical devices such as the processor, disk drives and monitor. On a stand-alone computer, software is divided into two categories: system software and application software. **System software** essentially comprises the computer's operating system (Topic O2) and any other programs required to run applications, while **application software** is installed by the user for specific purposes (e.g. word processing, image analysis, etc.). On networked computers, programs can also be run remotely. The same applies to computers attached to the Internet.

Programs and programming languages

Computer programs can be written in a variety of **programming languages**, which are conventionally described in terms of three levels. The first level is **machine code**, which is the binary code used by the computer's own processor. The second level includes a number of languages known as **assembly languages**. The third and subsequent levels are grouped as **higher-level languages** and include widely used programming languages such as Pascal and C. Programs written in assembly or higher-level programming languages must be converted into machine code before they will run. For assembly languages, this process is known as **assembly**, and for higher-level languages it is known as **compilation**. In Windows, files in machine code are known as **executable files** and have the extension .exe. There are no rules or conventions for naming such files in UNIX systems, and they are known as **executable images**. These files can be created and stored in the computer's memory until the operating system is told to run them. Alternatively, assembly or compilation can be carried out 'on the fly' if programs are executed remotely, for example over the Internet.

Scripts and scripting languages

Executable files (Windows) and executable images (UNIX systems) are written in machine code and are run by the computer's processor. Other program files are designed to be executed by another program, and such files are known as **scripts**. There are a variety of **scripting languages** that can be used, including Microsoft Visual Basic, JavaScript and PERL. Script languages are easier work with than compiled languages, but take longer to process, so they are ideal for short programs.

Popular languages in bioinformatics

A number of programming, scripting and markup languages (markup languages are discussed in Topic O4) are popular with bioinformaticists because they are versatile and can integrate a wide variety of types of data either in a stand-alone environment or over the Internet. Some of these languages are discussed below:

HTML and JavaScript
HTML is **hypertext markup language**, a language used to specify the appearance of a hypertext document, including the positions of hyperlinks. Since HTML is not a programming language, basic hypertext documents are static. **JavaScript** is a popular scripting language that adds to the functionality of hypertext documents, allowing web pages to include such features as pop-up windows, animations and objects that change in appearance when the mouse cursor moves over them.

Java
Java is a versatile and portable programming language that is designed to generate applications that can run on all hardware platforms, from large servers to individual PCs, without modification. The Java source code is based on C++ and can be run in a stand-alone fashion or from within a hypertext document, in

which case it is called an **applet** (small application). When executed, a Java program is converted into an intermediate language called **bytecode**, which is compiled into machine code as the program runs. Browsers must incorporate a Java plug-in interpreter called **Java Virtual Machine** for this purpose. Java applets may take a long time to download but the performance of the applet is not dictated by activities of the server. Java is a full programming language and is not the same as JavaScript, which is a scripting language. The names are similar because both languages use a similar syntax. As discussed above, JavaScript is used primarily to enhance World Wide Web (WWW) pages, while Java has a much broader scope.

PERL

PERL (Practical Extraction and Reporting Language) is a versatile scripting language, which is widely used in bioinformatics for applications such as the analysis of sequence data. PERL is a free product, providing compatibility with Windows, UNIX or other operating systems. It has excellent facilities for file handling and uploading and downloading files over the WWW.

XML

XML stands for **extensible markup language**. This is a new standard markup language that allows files to be described in terms of the types of data they contain. As a replacement for HTML, XML has the advantage of controlling not only how data are displayed on a WWW page, but also how the data is processed by another program or by a database management system.

Running programs over the Internet

Software does not have to be downloaded and installed on local computers but can be run over the Internet. This can be achieved in two ways. If the programs are **client-side**, they are supplied for example as JavaScript or Java applets that are embedded in HTML within a hypertext document. The utility of these programs might be limited by the capacity of the local machine. Furthermore, although both Internet Explorer and Netscape Navigator support JavaScript, the script is interpreted in slightly different ways by the two browsers. There is currently no clean solution to this problem. The alternative is to use **common gateway interface (CGI)** programs or **Java servlets**, in which case the software is run on the server itself (the programs are **server side**). Server side programs can be written in machine code or in a scripting language such as PERL or Java. It is easy to detect whether the software is running on a server because the URL will typically end with the extension .cgi. The performance of CGI programs is dependent on the number of current users (the server load). Some servers avoid bottlenecks by carrying out client instructions (e.g. homology searches) in their own time and then e-mailing the results to the client.

02 COMPUTER OPERATING SYSTEMS

Key Notes

Operating systems

A computer's operating system is a master program that links the processor to peripheral hardware items and allows software applications to run. The low-level operating system, usually provided as firmware, controls the basic functions of the computer, such as booting up, reading and writing to disks, responding to the keyboard and displaying information on the monitor. Higher-level operating systems generally provide a GUI that allows more complex interaction with the user.

Windows

Windows, marketed by Microsoft Corporation, is the most popular operating system for home and office PCs. Many home computers run on Windows 95/98/Me, while networked computers run on Windows NT/2000. The most recent version of this system, Windows XP comes in both personal and network versions. The low-level operating system is based on MS-DOS but incorporates the Windows GUI as an integral component. Files can be accessed using either the GUI or through the low-level operating system using the MS-DOS prompt.

UNIX

Most commercial workstations and servers run under variations of UNIX, a multi-user operating system developed in the 1970s. UNIX is not owned by any of the large computer companies and is largely in the public domain. It has been extensively modified and improved, and many different versions are available. Freeware variants such as GNU and LINUX have been widely adopted in the academic community. UNIX systems have no integral GUI, but many different GUIs are available, including GNOME, KDE and CDE.

Other operating systems

VMS is a multi-user operating system developed for computers made by the Digital Equipment Corporation, but this is increasingly being superseded by UNIX. Apple Macintosh computers have their own operating system, MacOS, which like Windows has an integral GUI.

Related topics

Running computer software (O1)
Software downloading and installation (O3)

Database management (O4)

Operating systems

On modern computers, the **operating system** is a master program that manages all peripheral hardware (e.g. monitor, keyboard, disk drives) and allows other **software applications** to run. There is a low-level operating system, sometimes called the **BIOS (Basic Input–Output System)**, which is largely or entirely in **firmware** (i.e. software stored in read-only memory). The BIOS handles activities such as deciding what to do when the computer is switched on after a cold start, reading and writing to disks, responding to input, displaying readable characters on the monitor and producing diagnostics. The higher-level operating system then takes over, and the computer acquires a typical **graphical user interface (GUI)** such as Windows. Files that contain instructions for the operating system are called **batch files** in Windows and **shell scripts** in UNIX systems. For example, the Windows batch file AUTOEXEC.BAT is required to initialize the disk operating system when you switch on your PC (see below).

Windows

Windows is the most familiar operating system on home and office PCs, and is wholly owned by Microsoft Corporation. Most stand-alone PCs currently run on **Windows 95** or **Windows 98** (often grouped as **Windows 9x**) or **Windows Me**. These operating systems are derived from the earlier **Microsoft Disk Operating System (MS-DOS)** and a GUI simply called **Windows**. From the launch of Windows 95, the GUI was integrated into the operating system and opens automatically when the operating system is loaded. Plain text files can be viewed without the benefit of the GUI using the **DOS shell** (a **shell** is an interactive interface between the user and an operating system, i.e. the part of the program that interprets and executes user commands). The DOS shell can be accessed from Windows by selecting Start, Programs, **MS-DOS Prompt**. MS-DOS and early versions of Windows were designed to run on stand-alone PCs. Now networks of computers are commonplace, Windows has been developed as a multi-user operating system. In **Windows NT** and **Windows 2000**, different users can have access to both personal and common files, which may all be located on a central server. The latest version of Windows, **Windows XP**, is available tailored for either home (stand-alone) or business (network compatible) use.

UNIX

Although Windows is the most popular operating system on PCs, most commercial workstations and servers run under variations of an operating system called **UNIX**. Unlike Windows, UNIX is not owned by any of the large computer companies, and since it is written in the standard programming language C, it has been modified and improved by many individuals, academic instructions and commercial companies for specific applications. There have been several public domain releases of operating systems that conform to the UNIX standard, such as **GNU** and **LINUX**. In particular, LINUX has become very popular in the scientific community. LINUX can be downloaded from the Internet or purchased at a nominal charge from one of several distributors. There are numerous GUIs for UNIX-like systems, which can be made to look like the familiar Windows or MacOS desktops. These include **GNOME** (GNU network object model environment), **KDE** (K desktop environment) and **CDE** (common desktop environment).

Other operating systems

Some older servers use the **VMS operating system** from the Digital Equipment Corporation (DEC). Apple Macintosh computers have their own operating system called **MacOS**, which has its own GUI. There is no simple way to view files on an Apple Macintosh without using the GUI. Other operating systems include OS/390, OS/400 and z/OS, which are used on some IBM computers.

O3 SOFTWARE DOWNLOADING AND INSTALLATION

Key Notes

Downloading methods	Computer software is downloaded when it is copied from a remote source onto a local computer hard drive. Programs can be downloaded directly from hypertext documents or by a file transfer protocol. Alternatively, programs may be received by e-mail.
Downloading from a hypertext document	Computer programs and other files may be embedded in a hypertext document, in which case they can be downloaded by clicking the appropriate hyperlink. This is often the most convenient way to download software.
Downloading from an FTP server	FTP is a standard way of exchanging files over the Internet. Many FTP servers allow anonymous FTP, where users can log in remotely, search through file directories and download any files they wish.
Receiving programs by e-mail	Files can also be sent and received as e-mail attachments, which can be downloaded from the mail server onto a local computer. In some cases, a local computer encountering an unrecognized file format may assume that it is a text file.
Archiving and compression	Files for downloading are often compressed to save space and limit downloading time. Also, compression allows files to be grouped together in archives. In Windows and MacOS, archiving and compression are carried out together using utilities such as WinZip or Stuffit. In UNIX, archiving and compression are separate tasks.
Installation methods	Before a downloaded program can be executed, it must be uncompressed and extracted from an archive. This is achieved with utilities such as Stuffit Expander, PkUnzip or UUencode. The program must then be installed on the computer so it can be recognized by the operating system and interpreted by the processor. The conventions for achieving this differ between Windows, MacOS and UNIX environments.

Related topics: Running computer software (O1)   Computer operating systems (O2)

Downloading methods

Computer software is often obtained on media such as floppy disks or CDs. CDs are the most versatile, not just because of their capacity (650 MB for a standard 72-min disk, and even more with data compression), but also because the standard file system, known as **ISO 9660**, is interpreted by all major computer platforms including Windows, MacOS and UNIX. However, it is often more convenient to download programs and other files from a local computer network or from the Internet. A file is **downloaded** when it is copied from a remote source onto a local computer. Conversely, a file is **uploaded** when it is copied from a computer's hard drive to a remote source. Downloading from the Internet can be achieved in three ways:

- Files can be downloaded directly from a hypertext document.
- Files can be obtained from an FTP server.
- Files can be received by e-mail.

Downloading from a hypertext document

This is the simplest method and is usually a case of clicking the appropriate hypertext link. In many cases, the browser will automatically detect the embedded file type and initiate a download procedure. The user on the local computer is allowed to specify the destination of the downloaded file and its name, if the default name (from the server) is not appropriate or will overwrite an existing file. In other cases, clicking may attempt to open the file remotely, so it may be necessary to right click (on PCs) or hold (on Apple Macs) the link with the mouse and choose 'save as' to download the file. It is strongly recommended to have virus-checking software installed and running if you are performing regular downloads!

Downloading from an FTP server

FTP stands for **file transfer protocol** and is a standard system for exchanging files over the Internet. FTP servers may restrict access to authorized individuals, who have to enter a username and password to gain access to the files. However, in many cases access is free, and users can log in anonymously and download whatever they like. This is known as **anonymous FTP**.

FTP servers host sites that contain lists of files arranged in a hierarchical directory. Files for public access are often located in a directory called 'public' or 'pub'. Some dedicated search tools, such as **Archie**, are available to locate specific files by searching directories of FTP sites. Otherwise, standard World Wide Web (WWW) **search engines** can be used to locate files (see Topic A3). Note that for FTP sites running on UNIX servers, the file names are case sensitive. This is not the case for sites based on a Windows operating system or MacOS.

Nowadays, anonymous access to FTP sites is usually achieved by typing the name of the FTP server in an Internet browser address window or clicking on an appropriate link. Both Internet Explorer and Netscape Navigator support anonymous FTP. Such sites have addresses with the configuration ftp.name.org or similar, although the user may be required to type **ftp://ftp.name.org** in the address bar of the browser rather than using the standard hypertext transfer protocol (http://......). Alternatively, a dedicated FTP program may be used to access the files. Such programs may have either a text or graphical user interface. Once the FTP site has been accessed, the anonymous user may be asked to log in and give a password, which should be 'anonymous' and an e-mail address respectively. The following UNIX-style commands are then used to access and download files:

ls	List directory of files on remote server
cd pub	Change directory to public files
cd pub/sequence	Change directory to public files in the directory 'sequences'
get sequence.gz	Download a file called sequence.gz (the suffix is explained below)
bye	Log off

FTP has some limited help facilities which can be accessed using the commands 'help' or '?'. The following commands are useful if large files are being downloaded:

prompt   Do not prompt for multiple access
verbose   Do not display output on the screen
mget *.gz   Multiple get all files with the specified name (* is used as a wildcard)
bg   Put the job in the background

Receiving programs by e-mail

Programs and other files can also be received by e-mail. This is achieved by appending the file to the e-mail message as an **attachment**, which can be downloaded (saved) by the recipient. In many cases, the format of the attached file will be recognized automatically. However, with some local mail clients, the original format of the file may not be recognized and will be interpreted by default as a text file, in which case a program such as Microsoft Word is opened automatically and tries to read and present the data as text. This is generally successful for text documents but for other files, for example executable programs, images and other multimedia objects, the result is a random mix of letters, numbers and other characters.

Archiving and compression

Most FTP sites store files in a **compressed format**, which reduces the file sizes (and therefore speeds up downloading) and also allows many files to be compressed together in a single entity called an **archive**. Different computer platforms have different compression systems, which are revealed by the file suffix. For Windows systems, files are usually compressed using **PkZip** or **WinZip**, and have the extension .zip. For Apple Macintosh systems, the programs **Stuffit** and **DropStuff** are widely used. The compression formats include **BinHex** (suffix .hqx), **Stuffit** (suffix .sit) and **self-extracting archive** (suffix .sea). UNIX systems support **UNIX compress** and **gzip** (suffix .Z or .gz) for compression and **tar** (suffix .tar) for archiving.

Installation methods

One a program has been downloaded it must be installed and run on the local computer. If the file or file archive was compressed, it must first be **uncompressed** (expanded, unzipped, uncrunched, etc.) using utilities such as **PkUnzip** or **Stuffit Expander**. There are some very useful programs that allow cross-platform archiving and compression/uncompression. For example, **UUencode** will convert binary files from Windows, MacOS or UNIX platforms into ASCII format and allow their expansion on any of these machine types.

Once a program file is uncompressed it can be installed. In a Windows or MacOS environment, this can be achieved using the operating system utilities for installation, or the expanded archive may include a file with the name install.exe or start.exe, which must be run from the desktop.

UNIX programs are usually distributed as source code because the C compiler and other programming utilities are part of the UNIX standard. If the downloaded file is called 'package.tar.gz' or 'package.tar.Z', the first job is to uncompress it using the instructions 'gunzip package.tar.gz' or 'uncompress package.tar.Z', and then to unpack the archive with 'tar xf package.tar'. This will recreate the file system. In a fairly simple case the new file system might contain a file called Makefile and the package should then be compiled with instructions such as:

```
make all
make test
make clean
make install
```

With very simple cases, 'make' alone should suffice.

More complex packages may not have a Makefile and the authors may have written a script conventionally called **configure**. In this the user types './configure' to run the script and it will produce some diagnostics. The purpose of configure is to explore the local operating system, find out what utilities are available and already installed and (assuming there is no problem) to create a Makefile.

04 DATABASE MANAGEMENT

Key Notes

Flat files and markup languages	Biological data is generally stored as text in flat files. Flat files are text files that lack any form of markup language, the hidden instructions that dictate how text is displayed on screen and in printed documents. Markup tags are transparent to many bioinformatics software applications, such as sequence analysis programs.
Databases	Computer databases are collections of records (individual entries) each comprising data in the same set of fields (categories). Records may be maintained as separate files, or may be combined into a single file. The database is one component of a DBMS, a suite of programs for accessing, searching and amending the records in the database. In relational databases, the data are organized in tables and searched using a standard language called SQL. In object-orientated databases, data are defined as objects, which may include text, images, videos and audio files. The organization of object-oriented databases is more flexible than that of relational databases allowing more complex relationships between data to be modeled. Databases incorporating both relational and object-oriented programming concepts are widely used in bioinformatics.
Related topics	File formats (C1) Genome and organism-specific Annotated sequence databases (C2) databases (C3) Miscellaneous databases (C4)

Flat files and markup languages

On a computer system, a **file** is a discrete collection of bytes that can be manipulated (moved, copied, deleted etc.) as a single entity. A file may either constitute a **program** (Topic O1) or a data file that is processed by a program (e.g. a document that can be read by Microsoft Word). In the context of bioinformatics, files are used to store structured biological data. Most raw biological data can be stored in the form of **text**, for example nucleotide and protein sequences, protein structural coordinates and matrices of gene expression profiles. **Text files** can be handled by various software applications such as **text editors** (e.g. SimpleText), **Internet browsers** (e.g. Internet Explorer, Netscape Navigator) and **word processor applications** (e.g. Microsoft Word, Corel WordPerfect). Other types of biological data are stored as images, for example gene expression patterns and pictures of two-dimensional-protein gels. In some cases, the raw data in the images are converted into numbers that can be stored in text files. For example, this is the case for microarray image data (Topic J1).

Most software applications that handle text include a **markup language** that specifies how the text should be displayed on screen or in a printed document. These instructions comprise hidden character sets, known as **tags**. In Microsoft Word, for example, the markup language controls the font, size, color, paragraph structure etc. of the text. Other familiar markup languages include **HTML** (**hypertext markup language**), which controls the display of text on WWW pages and enables hyperlinks to be inserted, and **XML (extensible markup language)**, which allows the integral description of data objects (see Topic O1). Such tags are often transparent, however, if the text is used by another software application, such as a sequence analysis program. Therefore, it is best to save biological data files in a simple format with no markup language. These text-only files are known as **flat files**. Text editor programs such as **SimpleText** and **NotePad** are suitable for handling flat files, and flat files can be generated in word processor programs such as Microsoft Word by saving as **text only**. Standard flat file formats for the presentation of sequence and structural data are discussed in Topic C1.

Databases

A **database** is a collection of structured information, often stored in the form of flat files in the case of bioinformatics data. Individual database entries are known as **records** and each record comprises the same set of **fields** (categories of data). For example, in a sequence database such as GenBank, each record represents a deposited sequence and fields include accession number, sequence name, taxonomy of source species, literature references, and the sequence itself (Topic C2). Computer databases are usually associated with software that allows the information to be accessed, amended and searched. This software is known as a **database management system (DBMS)** and also controls the security and integrity of the data. Searches are made possible by indexing the records, which in the case of flat files is achieved by looking for text strings in particular fields. In the case of a sequence database, the accession number could be used for index purposes. Several different types of database are used in bioinformatics.

A **relational database** is organized as tables, each table comprising a group of records (also known as **tuples**) with the same fields (known as **attributes**). This allows related data to be linked (reassembled) as required without reorganizing the original tables. For example, a sequence table might contain records with the attributes *accession number* and *protein sequence*, while a function table might contain records with the attributes *accession number* and *protein function*. Matching attributes in different tables can be joined to bring together related records, in this case linking *protein sequence* and *protein function*. The industry-standard language used to interrogate and process data in a relational databases is **SQL (symbolic query language)**. This is incorporated into familiar and widely used relational database management systems such as **Microsoft Access** and **Oracle**.

An **object-oriented database** has a more flexible organization, that is, it does not depend on the formal 'table, row and column' format of relational databases. Data are defined as **objects**, which have a **class hierarchy**, that is, they can be grouped into classes and subclasses etc. in a hierarchical manner. Properties attributable to classes of objects are inherited through the hierarchy; these may be general in the upper levels of the hierarchy but may become more specialized in the lower levels. Properties or procedures attributable to data objects are known as **methods**. The flexibility of the data organization in object-orientated databases allows more complex relationships between datasets to be modeled than is possible with relational databases. Object-oriented database management systems are also capable of handling multimedia objects (pictures, videos and audio files) while relational DBMSs are often restricted to numbers, alphanumeric text and dates. Pure object-oriented DBMSs include **Object Store** and **ONTOS DB**. ACeDB (A *C. elegans* database; Topic C3) is an example of a customized object-oriented database. It is more common to see bioinformatics databases incorporating relational DBMS features in an object-oriented programming environment. Such **object-relational DBMSs** are generally accessed using a language based on SQL.

Developments in object-oriented programming have led to attempts to have object definitions that are common across different computer systems. This is useful for the integration of **distributed databases**, that is, databases that are physically stored on two or more separate computer systems. An interface definition called **CORBA (Common Object Request Brokering Architecture)** has been developed which can be used to integrate large distributed bioinformatics databases. Software such as CORBA that functions as a conversion or translation layer in distributed systems is sometimes called **middleware**.

FURTHER READING

General reading   Attwood, T.K., Parry-Smith, D.J. (1999) *Introduction to Bioinformatics*. Addison Wesley Longman, Harlow, Essex.

Lesk, A.M. (2002) *Introduction to Bioinformatics*. OUP, Oxford.

More advanced reading

Section A   Boguski, M.S. (1998) Bioinformatics – a new era. *Bioinformatics: A Trends Guide* **5**, 1–3.

Brownstein, M.J., Trent, J.M. and Boguski, M.S. (1998) Functional genomics. *Bioinformatics: A Trends Guide* **5**, 27–29.

Davidson, D. and Baldock, R. (2001) Bioinformatics beyond sequence: mapping gene function in the embryo. *Nature Rev. Genet.* **2**, 409–417.

Goodman, N. (2001) Biological data becomes computer literate: new advances in bioinformatics. *Curr. Opin. Biotechnol.* **13**, 68–71.

Hales, M. (1997) Basic Internet facilities. *Trends Guide to the Internet* 8–9.

Patsan S (1997) Evolution of the Net. *Trends Guide to the Internet* 5.

Section B   Brown, T.A. (1994) *DNA Sequencing: The Basics*. IRL Press, Oxford.

Duggan D.J., Bittner M., Chen Y., Meltzer P., Trent J. (1999) Expression profiling using cDNA microarrays. *Nature Genet.* **21**(Suppl.), 10–14.

Lander, E.S. (1999) Array of hope. *Nature Genet.* **21**(Suppl.), 3–4.

Lipshutz, R.J., Fodor, S.P.A., Gingeras, T.R. and Lockhart, D.J. (1999) High density synthetic oligonuceotide arrays. *Nature Genet.* **21**(Suppl.), 20–24.

Phizicky, E.M. and Fields, S. (1995) Protein–protein interactions: methods for detection and analysis. *Microbiol Rev.* **59**, 94–123.

Section C   Kanehisa, M. (1998) Databases of biological information. *Bioinformatics: A Trends Guide* **5**, 24–26.

Section D   Lewitter, F. (1998) Text-based database searching. *Bioinformatics: A Trends Guide* **5**, 3–5.

Section E   Altschul, S.F. (1998) Fundamentals of database searching. *Bioinformatics: A Trends Guide* **5**, 7–9.

Brenner, S.E. (1998) Practical database searching. *Bioinformatics: A Trends Guide* **5**, 9–12.

Patthy, L. (1999) *Protein Evolution*. Blackwell Science Ltd, Oxford.

Section F   Eddy, S.R. (1998) Multiple-alignment and sequence searches. *Bioinformatics: A Trends Guide* **5**, 15–18.

Section G   Lake, J.A. and Moore, J.E. (1998) Phylogenetic analysis and comparative genomics. *Bioinformatics: A Trends Guide* **5**, 22–23.

Section H   Burge, C.B. and Karlin, S. (1998) Finding the genes in genomic DNA. *Curr. Opin. Struct. Biol.* **8**, 346–354.

Gaasterland, T. and Oprea, M. (2001) Whole-genome analysis: annotations and updates. *Curr. Opin. Struct. Biol.* **11**, 377–381.

Haussler, D. (1998) Computational genefinding. *Bioinformatics: A Trends Guide* **5**, 12–15.

Lewis, S., Ashburner, M. and Reese, M.G. (2000) Annotating eukaryote genomes. *Curr. Opin. Struct. Biol.* **10**, 349–354.

Rogic, S., Mackworth, A.K. and Oulette, F.B.F. (2001) Evaluation of gene-finding programs on mammalian sequences. *Genome Res* **11**, 817 832.

Stein, L. (2001) Genome annotation: from sequence to biology. *Nature Rev. Genet.* **2**, 493–503.

Section I

Hofmann, K. (1998) Protein classification and functional assignment. *Bioinformatics: A Trends Guide* **5**, 18–21.

Lesk, A.M. (2002) *Introduction to Bioinformatics*. OUP, Oxford.

Patthy, L. (1999) *Protein Evolution*. Blackwell Science Ltd, Oxford.

Section J

Altman, R.B. and Raychaudhuri, S. (2001) Whole-genome expression analysis: challenges beyond clustering. *Curr. Opin. Struct. Biol.* **11**, 340–347.

Bassett, D.E., Eisen, M.B. and Bogusky, M.S. (1999) Gene expression informatics – it's all in your mine. *Nature Genet.* **21**(Suppl.), 51–55.

Bowtel, D.D.L. (1999) Options available – from start to finish – for obtaining expression data by microarray. *Nature Genet.* **21**(Suppl.), 25–32.

Hess, K.R., Zhang, W., Baggerly, K.A. *et al.* (2001) Microarrays: handling the deluge of data and extracting reliable information. *Trends Biotechnol.* **19**, 463–468.

Quackenbush, J. (2001) Computational analysis of microarray data. *Nature Rev. Genet.* **2**, 418–427.

Raychaudhuri, S., Sutphin, P.D., Chang, J.T. and Altman, R.B. (2001) Basic microarray analysis: grouping and feature reduction. *Trends Biotechnol.* **19**, 189–193.

Schena, M., Shalon, D., Davis, R.W. and Brown, O.P. (1995) Quantitative monitoring of gene expression patterns with a complementary DNA microarray. *Science* **270**, 467–470.

Velculescu, V.E., Vogelstein, B. and Kinzler, K.W. (2000) Analysing uncharted transcriptomes with SAGE. *Trends Genet.* **16**, 423–425.

Velculescu, V.E., Zhang, L., Vogelstein, B. and Kinzler, K.W. (1995) Serial analysis of gene expression. *Science* **270**, 484–488.

Section K

Andersen, J.S. and Mann, M (2000) Functional genomics by mass spectrometry. *FEBS Lett.* **480**, 25–31.

Beavis, R.C. and Fenyö, D. (2000) Database searching with mass-spectrometric information. *Proteomics: A Trends Guide* **1**, 22–27.

Choudhary, J.S., Blackstock, W.P., Creasy, D. and Cottrell, J.S. (2001) Matching peptide mass spectra to EST and genomic DNA databases. *Trends Biotechnol.* **19**(Suppl.), 17–22.

Crawford, M.E., Cusick, M.E. and Garrels, J.I. (2000) Databases and knowledge resources for proteomics research. *Proteomics: A Trends Guide* **1**, 17–21.

Görg, G. (2000) Advances in 2D gel techniques. *Proteomics: A Trends Guide* **1**, 3–6.

Lee, K.H. (2001) Proteomics: a technology-driven and technology-limited discovery science. *Trends Biotechnol.* **19**, 217–222.

Schweitzer, B. and Kingsmore, S.F. (2002) Measuring proteins on microarrays. *Curr. Opin. Biotechnol.* **13**, 14–19.

Templin, M.F., Stoll, D., Schrenk, M., Traub, P.C., Vohringer, C.F. and Joos, T.O. (2002) Protein microarray technology. *Trends Biotechnol.* **20**, 160–166.

Section L

Boulton, S.J., Vincent, S. and Vidal, M. (2001) Use of protein-interaction maps to formulate biological questions. *Curr. Opin. Chem. Biol.* **5**, 57–62.

Hodgman, C., Goryanin, I. and Juty, N. (2001) Reconstructing whole-cell models. *Genomics* **2**(Suppl.), 109–112.

Ito, T., Chiba, T. and Yoshida, M. (2001) Exploring the protein interactome using comprehensive two-hybrid projects. *Trends Biotechnol.* **19**(Suppl.), 23–27.

Pelletier, J. and Sidhu, S. (2001) Mapping protein–protein interactions with combinatorial biology methods. *Curr. Opin. Biotechnol.* **12**, 340–347.

Uetz, P. (2001) Two-hybrid arrays. *Curr. Opin. Chem. Biol.* **6**, 57–62.

Xenarios, I., Eisenberg, D. (2001) Protein interaction databases. *Curr. Opin. Biotechnol.* **12**, 334–339.

Section M

Bailey, D.S., Furness, M. and Dean, P.M. (1999) New tools for quantifying molecular diversity. *Pharmainformatics: A Trends Guide* **4**, 6–9.

Section N

Brocklehurst, S.M., Hardman, C.H. and Johnston, S.J.T. (1999) Creating integrated computer systems for target discovery and drug discovery. *Pharmainformatics: A Trends Guide* **4**, 12–15.

Gardner, S.P. and Flores, T.P. (1999) Integrating information technology with pharmaceutical discovery and development. *Pharmainformatics: A Trends Guide* **4**, 2–5.

Gund, P. and Sigal, N.H. (1999) Applying informatics systems to high-throughput screening and analysis. *Pharmainformatics: A Trends Guide* **4**, 25–29.

Mason, J.S. (1999) Computational screening: large-scale drug discovery. *Pharmainformatics: A Trends Guide* **4**, 34–36.

Roberts, B.R. (2000) Screening informatics: adding value with metadata structures and visualization tools. *High-Throughput Screening 1: A Supplement to Drug Discovery Today* **2**, 10–14.

Terfloth, L. and Gasteiger, J. (2001) Neural networks and genetic algorithms in drug design. *Genomics* **2**(Suppl.), 102–108.

GLOSSARY

2D-PAGE	Two-dimensional gel electrophoresis. A technique for separating complex mixtures of proteins according to isoelectric point and mass.
ab initio	Describing an analysis method carried out from first principles. Applied to gene prediction algorithms that do not incorporate homology searching and structural modeling algorithms that do not rely on the structures of known proteins.
affine gap penalty	See *gap penalty*.
algorithm	A set of rules for calculating or problem solving carried out by a computer program.
alignment	Arrangement of two or more nucleotides of protein sequences to maximize the number of matching monomers.
alignment score	A numerical value that describes the quality of a sequence alignment (q.v.). Also see *percent identity, percent similarity*.
analog	A protein which shares the *fold* (q.v.) of another protein, but for which other evidence of divergence from a common ancestor is weak (see also *homolog*).
annotation	(1) Finding genes and other important elements in raw sequence data (structural annotation). (2) Determining the function of genes and proteins (functional annotation).
anonymous FTP	See *FTP*.
applet	A small application program that can be executed by selecting a link or that executes automatically when a *HTML* document is loaded by a *browser* (q.v.). Applets are often written in *Java* (q.v.) allowing them to be run on different operating systems and use the *GUI* (q.v.) to provide input/output with the user.
application	Software installed on a computer for a specific purpose (e.g. word-processing) as opposed to the system software or *operating system* (q.v.).
archive	A collection of files. These may be stored in a single *compressed* file (q.v.) in a special archive format. Also see *zip, gz*.
array	(1) Computational equivalent of a table but can have more than two dimensions. (2) See *matrix*. (3) See *microarray*.
ASCII	American Standard Code for Information Interchange. A code using single *bytes* (q.v.) to represent characters, with codes 0 to 127 (in decimal) used for the Latin alphabet, Arabic numerals, punctuation etc. ASCII codes below 32 (in decimal) are not normally printable characters on most computers, but represent codes for terminal-input, cursor and printer control.
assembly	(1) The process of joining individual DNA sequences to generate larger *contigs* (q.v.). (2) The conversion of a second-level programming language (assembly language) into *machine code* (q.v.).
attachment	A file that is sent appended to an e-mail.
B-series	*N*-terminal fragments in a peptide ladder generated by *mass spectrometry* (q.v.).

base calling   Determining the sequence of bases in a DNA molecule from *trace data* (q.v.).

batch   On a computer, a job running in the background while the user carries out a different task.

batch file   A command file (a file that gives instructions to the operating system) in a Windows environment. Cf. *shell script*.

bioinformatics   The branch of information technology that deals with the storage and analysis of biological data.

BIOS   Basic Input–Output System. The lowest-level *operating system* (q.v.) of a computer, controlling elementary functions such as reading and writing to disks, responding to input and displaying readable characters on the monitor. Normally required to run the *operating system* (q.v.).

bit   Binary digit. A bit can represent one of two values (0 and 1 – sometimes referred to as 'off' or 'on').

BLAST   Basic Local Alignment Search Tool. A program for sequence database similarity searching (see also *FASTA*).

BLOSUM matrices   *Substitution score matrices* (q.v.) for proteins.

Boolean network   A network with one or more inputs, whose outputs are constrained by an explicit set of logical rules.

bootstrapping   A statistical method used to evaluate the reliability of phylogenetic trees.

browser   An application that processes documents written in *HTML* (q.v.) and interprets the markup language to display formatted *web pages* (q.v.).

byte   Eight bits. Thus a byte can have values ranging from 0000000 (0) to 11111111 (255). A character (letter, numeral etc., see *ASCII*) is usually represented by one byte.

C   A procedural compiled *programming language* (q.v.) widely used for bioinformatics software. C is part of the UNIX standard.

C++   An *object-oriented programming language* (q.v.) compatible with *C*.

cDNA   Complementary DNA, reverse transcribed from messenger RNA.

CGI   Common gateway interface. A program that can be used over the Internet but that runs *server-side* (q.v.) rather than *client-side* (q.v.) and whose performance thus depends on the server load.

cheminformatics   The branch of information science that deals with chemistry data.

chirality   Describing molecules with the same structure but occurring in multiple forms due to the presence of one or more asymmetric atoms. See *enantiomer, diastereoisomer*.

cladogram   A dendrogram in which each node has two branches, representing evolutionary history as speciation by bifurcation of the evolutionary lineage (cladogenesis).

client-side   A client-side program runs within a browser, rather than remotely at the server (cf. *server-side, CGI*). Such programs may require input over the Internet, but will run irrespective of the processor and operating system on the local machine. They are generally *Java applets* or *JavaScript* (q.v.).

command language   The language used for giving instructions to a computer operating system.

comparative modeling	The process of predicting protein structure based on related sequences of known structure.
compile	Convert a higher-level programming language into *machine code* (q.v.).
compression	Making a computer file smaller to reduce transfer time over a network. Also see *archive, zip, gz.*
content	In gene prediction, an extended and variable region of DNA sequence with no particular motifs but certain conserved characteristics.
contig	A collection of individual DNA sequence reads assembled into a larger contiguous sequence.
CORBA	Common Object Request Brokering Architecture. A request brokering system, acting as a conversion or translation layer in distributed databases.
database	On a computer, a collection of data records either in a single file or as multiple files. The central component of a *database management system* (q.v.).
database management system (DBMS)	A software suite including a *database* (q.v.) and the utilities required to organize, search, and update it, maintain data security and control access.
DBGET/LinkDB	On-line data retrieval tool developed by the Institute for Chemical Research and the Human Genome Center in Japan. See also *Entrez, SRS.*
deletion	Part of a sequence alignment where one sequence appears to have fewer (deleted) monomers compared with another sequence (see also *insertion* and *gap*).
dendrogram	A branching graph used to represent e.g. phylogenetic relationships or the clustering of microarray data. See *cladogram.*
desktop	The front end of a *GUI* (q.v.).
diastereoisomers	Molecules with multiple chiral centers, which may exist in many forms with different physical and chemical properties. See *chirality.*
distributed database	A *database* (q.v.) comprising data and associated software distributed over a network or the *Internet* (q.v.). See also *COBRA, middleware.*
DL convention	A way of describing *enantiomers* (q.v.) based on a comparison of their structures to D-glyceraldehyde and L-alanine.
domain	Usually used to describe part of a protein that can fold and carry out a function independently (cf. *fold*), but sometimes used more generally to indicate part of a protein sequence, for instance a 'glycine-rich domain', or a geometrically distinct part of a protein structure.
DOS	Disk Operating System. A lower-level operating system used to read and copy files and carry out other basic functions. MS-DOS was developed by Microsoft now integrated with the *Windows GUI* (q.v.)
download	Transfer files from a computer network (or the *Internet,* q.v.) onto a local computer. Cf. *upload.*
dynamic programming	A method for the comparison and alignment of *strings* or *sequences* in a way that allows the computationally efficient incorporation of gaps.
E value	An expectation value used to test the significance of a sequence similarity score (see also *p value*)

enantiomers   Molecules of identical structure but different 'handedness'. Applies esp. to organic molecules with an asymmetric carbon atom. See *chirality, diastereoisomer*.

ESI   Electrospray ionization. A *mass spectrometry* (q.v.) technique suitable for the ionization of large molecules such as proteins without significant degradation. See also *MALDI*.

EST   Expressed sequence tag. A partial *cDNA* (q.v.) obtained by high-throughput, random single-pass sequencing of cDNA libraries.

Entrez   On-line data retrieval tool developed by the National Center for Biotechnology Information (NCBI). See also *DBGET/LinkDB, SRS*.

**executable file,   A file in machine code, which can be interpreted by the computer's processor
image**   under the control of the *operating system* (q.v.).

FASTA   (1) A sequence alignment algorithm. (2) FASTA format. A flat file format for the representation of sequence data.

flat file   A plain text *file* (q.v.). A file with no markup characters (see *markup language*).

file   On a computer, a discrete collection of *bytes* (q.v.) that can be manipulated as a single entity.

filter   See *mask*.

fold   The basic tertiary structure of a protein, including the secondary structure elements, their sequential connections and relative spatial positions.

fold recognition   Protein structure prediction through the detection of very remote structural relationships or *homologies* (q.v.).

FTP   File transfer protocol. A method of transferring files from one computer to another. Access can be restricted (requires user name and password) or unrestricted (anonymous FTP).

functional genomics The study of gene function on a global scale.

gap   A part of a sequence alignment where one sequence contains no aligned monomer. Can be interpreted in terms of evolutionary *insertion* (q.v.) and *deletion* (q.v.) of monomers.

gap penalty   A penalty term subtracted from a sequence similarity score to account for *gaps* (q.v.) in a sequence alignment.

GDE   A flat file format for sequence data, similar to *FASTA* (q.v.).

genetic algorithm   An artificial intelligence method and an example of a biological metaphor in computing. Essentially solutions to a problem or a theory are represented as *strings* and these are regarded as 'chromosomes' that replicate, mutate and recombine. Only the chromosomes that are 'fittest' (i.e. correspond to the better solutions) survive the iterations of the method.

GSS   Genome Survey Sequence. A random, single pass genomic DNA sequence (cf. *EST*).

GUI   Graphical user interface. A higher-level operating system with a more sophisticated user interface than the basic text-input style of a lower-level operating system. The Windows desktop is one example of a GUI.

gz
A file extension for UNIX files compressed by using the utility gzip (gunzip is the utility for the reverse process). See also *zip*.

heuristic
Of a computer program, making guesses to obtain approximate results but much faster than possible with exhaustive searching.

hidden Markov model (HMM)
A *Markov model* (q.v.) used in sequence analysis in which one or more variables is hidden.

hierarchical clustering
Any data clustering method in which the most similar data points are recursively clustered into a hierarchical tree. Alternatives include *k-means clustering* and *self-organizing maps* (q.v.).

HMM
See *hidden Markov model*.

homolog
A biological macromolecule related to another by divergent evolution from a common ancestor.

homology
An evolutionary relationship of two molecules deriving from a common ancestor.

HTML
Hypertext markup language. The *markup language* (q.v.) that controls how text and multimedia objects are displayed on *web pages* (q.v.).

HTTP
Hypertext transfer protocol. The protocol used to exchange information over the *World Wide Web* (q.v.).

hyperlink
A word or object in a *hypertext* document (q.v.), usually highlighted, which acts as a link to another document, either on the same computer or any other computer linked to the *Internet* (q.v.).

hypertext
Referring to a document (or part thereof) written in *HTML* (q.v.).

indel
See *gap*.

insertion
Part of a sequence alignment where one sequence appears to have extra (inserted) monomers compared with another sequence (see also *deletion, gap*).

Internet
An international computer network governed by a set of *protocols* (q.v.) called *TCP/IP* (q.v.).

Internet Explorer
A popular *browser* (q.v.) available from Microsoft. See also *Netscape Navigator*.

IP
Internet Protocol. See *TCP/IP*.

IP address
An address to which data is sent or from which it is received over the *Internet* (q.v.). Conventionally four integers separated by dots, e.g. 195.172.6.15. See also *URL*.

Java
An *object-oriented programming language* (q.v.) designed to run on most platforms. Java programs can be run alone or from within an HTML document, in which case they are supplied as an *applet* (q.v.).

JavaScript
A *scripting language* (q.v.) designed for web-based applications.

k-means clustering
A non-hierarchical data clustering method in which the number of clusters is defined before analysis begins.

KEGG
Kyoto Encyclopedia of Genes and Genomes. Excellent on-line resource for information on molecular pathways.

knockout	Gene disruption or deletion by homologous recombination.
LINUX	A UNIX-like operating system named after its inventor, Linus Thorvald.
lead compound	A substance that has many of the characteristics of an ideal drug and which interacts with a specific target. Can be modified and improved (lead optimization) to produce drug candidates with reduced side effects.
long branch attraction	The placement of groups of rapidly evolving sequences together in phylogenetic trees, irrespective of their true relationships.
low complexity segment	A sub-sequence of a protein or nucleic acid dominated in composition by a small number of monomer types.
machine code	The binary code interpreted by a computer's processor.
make	A UNIX utility for maintaining a hierarchy of files and commonly used for the maintenance of *source code* (q.v.). By default, make follows the instructions in a file called Makefile. It is a powerful utility as it keeps a note of files that are unchanged since the last make was performed and does not (for example) re-compile source code that is unchanged.
MALDI	Matrix-assisted laser desorption/ionization. A technique for generating ions in *mass spectrometry* (q.v.), which is suitable for the analysis of large proteins without significant degradation.
Markov model	A probabilistic statistical model used in sequence analysis where the probability of each nucleotide occurring is dependent on the nucleotides preceding it. Also see *hidden Markov model*.
markup language	A computer language that controls how text and graphics are displayed on the screen or in a printed document. See *XML, HTML*.
mask	A sequence is masked when a computer program ignores it to prevent spurious matches with many other sequences. Generally applies to repetitive DNA and vector sequences.
mass spectrometry (MS)	A technique for accurately measuring the mass/charge ratio of ions in a vacuum, and therefore calculating molecular masses.
matrix	(1) An experimental format, also known as an array, in which combinations of conditions are systematically tested in all possible pair-wise combinations. (2) Data arranged in rows and columns (as in gene expression matrix, distance matrix etc).
MIAME	Minimum information about a microarray experiment. A recent convention for the unambiguous presentation of *microarray* data (q.v.).
microarray	A miniature device, also known as a chip, containing hundreds or thousands of different molecules immobilized in a regular pattern. Typically **DNA microarray**, used for high throughput genotyping and expression analysis, and **protein microarray**, which is used for expression analysis and detecting protein interactions.
middleware	Software that acts as a translation layer in distributed systems operating on different platforms.
MIME type	Multiple Internet Mail Extensions. One of a variety of conventions for converting binary and other files into *ASCII* for transmission by e-mail. Also see *UUE*.

mirror	Identical web sites, hosted on different computers, such that the data might be acquired more quickly by users in specific countries.
motif	A short conserved region in a DNA sequence or protein. Cf. *fold*, *domain*.
ontology	Relationship between objects, especially in artificial intelligence systems.
open source	Software that can be downloaded and installed locally, modified and redistributed.
operating system	Master program on a computer that manages all hardware and allows application software to run. Often comprises low level *BIOS* (q.v.), a *disk-operating system* (q.v.) and higher level *GUI* (q.v.).
ORF	Open reading frame. The coding region of a gene, often split into exons and introns in eukaryotes.
ortholog	A *homolog* (q.v.) with identical function in a different organism (cf. *paralog*).
NBRF/PIR	National Biomedical Research Foundation/Protein Information Resource. A flat file format for representing sequence information.
Needleman–Wunsch	A *dynamic programming algorithm* (q.v.) for global sequence alignment (see also *Smith–Waterman*).
neighbor joining	A hierarchical clustering method, similar to *UPGMA* (q.v.), which is used in the Clustal W/X programs.
Netscape Navigator	A popular Internet *browser* (q.v.).
NMR	Nuclear magnetic resonance. The ability of atoms to switch between magnetic spin states in an applied magnetic field. NMR spectra depend on the type of atom, its chemical context and the proximity of atoms in space, and can therefore be used to determine protein structures. Cf. *X-ray crystallography*.
object-oriented (OO)	A programming method, incorporated, e.g., into object oriented databases, in which data are regarded as objects and associated with properties and routines known as methods. More flexible than procedural programming methods, therefore OO-databases allow more complex relationships between data sets to be modeled compared to *relational databases* (q.v.).
object-relational	Describing a *relational database* (q.v.) incorporating aspects of *object-orientated programming* (q.v.).
OMIM	On-Line Mendelian Inheritance in Man, an extensive on-line resource for human genetics and molecular biology.
p **value**	A probability used to test the significance of a sequence similarity score (see also *E value*).
PAM matrices	*Substitution score matrices* (q.v.) for proteins.
paralog	A *homolog* (q.v.) in the same genome with a different biological function (cf. *ortholog*).
Pascal	Procedural *programming language* (q.v.), named after Blaise Pascal and designed for teaching computing. The language has been largely superseded by *C* (q.v.).
PDB	(1) Protein Databank, the universal repository for protein structural data obtained by *X-ray crystallography* (q.v.) or *NMR spectroscopy* (q.v.). (2) The standard flat file format for protein structural data.

peptide mass fingerprinting	A method for protein annotation in which the masses of peptides (produced by protease digestion) are determined by *mass spectrometry* (q.v.) and used to search protein databases for matches to 'virtually digested' proteins.
percent identity	An *alignment score* (q.v.) based on the proportion of identical residues between two nucleotide or amino acid sequences. Cf. *percent similarity*.
percent similarity	An *alignment score* (q.v.) used for amino acid sequences in which a *substitution matrix* (q.v.) is used to rank the substitution scores of different amino acids.
PERL	Practical Extraction and Reporting Language. A versatile scripting language (see *script*), which is widely used in bioinformatics for applications such as the analysis of sequence data.
personalized medicine	Drugs that are targeted to individuals based on their genotype, helping to reduce adverse drug reactions due to differences in tolerance and metabolism (q.v. *pharmacogenomics*).
pharmacogenomics	The branch of biomedicine, dealing with the association between genome data and drug responses, that is hoped to make *personalized medicine* (q.v.) a reality.
pharmainformatics	The branch of information science that deals with handling biological and chemical data in the pharmaceutical industry.
pipeline	A sequence of experimental and bioinformatic procedures used to convert raw data into meaningful data, e.g. sequencing pipeline, annotation pipeline.
platform	A specific combination of computer hardware and *operating system* (q.v.).
portable	A program is said to be portable (platform independent) if it can be transferred across software and hardware architectures.
primary database	A database for primary sequence data. The primary nucleotide databases are NCBI GenBank, the European Molecular Biology Laboratory (EMBL) Nucleotide Sequence Database, and the DNA Database of Japan. The primary protein databases are SWISS-PROT and TrEMBL. See also *secondary database*.
probe	A molecule that is labeled and used to detect another molecule (target) through its specific affinity. Applies esp. to nucleic acids used to detect targets by hybridization and antibodies used to detect proteins.
programming language	A language used for writing computer programs. The first level of programming languages is *machine code* (q.v.). Other languages need to be converted into machine code before the program will run. See *assembly*, *compile*. Cf. *script*, *scripting language*.
proteome	The entire complement of proteins produced by a particular genome, including variants of the same basic protein generated by post-translational modification etc. The study of the proteome is known as proteomics.
protocol	A procedure for handling packages of data when computers communicate with each other.
PSI-BLAST	An iterative version of the *BLAST* (q.v.) algorithm.
QSAR	Quantitative structure–activity relationship. A mathematical function used to relate the structural features of a molecule to its biological function.
rational drug design	Drug design based on knowledge of protein structure and interactions with small ligands.

relational database	A *database* (q.v.) in which data records are organized as tables, allowing the data from tables containing similar fields to be linked together (related). Cf. *object orientated database*.
reverse engineering	Any modeling process based on the principle of developing the model to fit observed data (cf. *simulation*).
rooted tree	A phylogenetic tree in which the last common ancestor of all the species in the tree is present as an ancestral outgroup.
RS convention	Also known as CIP convention, a set of rules for defining the absolute configuration of chemical groups around a chiral center (see *enantiomer, chirality*). Cf. *DL convention*.
SAGE	Serial analysis of gene expression. A sequence sampling technique for transcriptional profiling in which small sequence tags are concatamerized, cloned and sequenced to determine relative levels of gene expression.
script	A sequence of instructions executed by another software application, not directly by the computer's processor. Scripts are written in a scripting language such as *PERL* (q.v.) or *Java* (q.v.).
search engine	A *server-side* (q.v.) facility that allows the *World Wide Web* (q.v.) to be searched for pages containing particular words, phrases or multimedia objects.
secondary database	A database of sequence information derived from the data in *primary databases* (q.v.). Examples include PROSITE, BLOCKS, Pfam and PRINTS.
server-side	Located on a web server rather than on the browser program being used as a client. See *CGI, client-side*.
signal	In gene prediction, a specific functional site in a DNA sequence that identifies a gene.
simulation	Any modeling process based on the principle of testing different models until one is found that fits the data (cf. *reverse engineering*).
Smith–Waterman	A *dynamic programming algorithm* (q.v.) for local sequence alignment (see also *Needleman–Wunsch*).
SNP	Single nucleotide polymorphism. A change in DNA sequence at a single residue. There are millions of SNPs in the human genome, which are thought to contribute to many complex phenotypes including drug response patterns.
SOM	Self-organizing map. A clustering method similar to *k-means clustering* (q.v.) but refined through the use of neural networks.
SRS	Sequence retrieval system. A data retrieval tool similar to *Entrez* (q.v.) and *DBGET/LinkDB* (q.v.) but it can be used as a stand-alone program as well as over the Internet.
shell	A text-based utility for giving commands to a computer, e.g. DOS-shell (q.v. *DOS*).
shell script	A command file (a file that gives instructions to the computer's operating system) in a UNIX environment. Cf. *batch file*.
SMILES	Simplified Molecular Input Line Entry Specification. A system for representing chemical structures using simple text *strings* (q.v.).

source code	The *ASCII* text of a *programming language* (q.v.).
SQL	Symbolic query language. The industry-standard language used to interrogate and process data in a *relational database* (q.v.).
string	One-dimensional *array* (q.v.) of objects of some kind, as in character string (a linear sequence of characters) or bit string (a string of bits in machine code).
structural alignment	Creating protein sequence alignments using structural information.
structural genomics	High-throughput determination of protein structure, perhaps involving structure determination for whole *proteomes* (q.v.).
substitution score matrix	A matrix containing scores for matching protein or nucleic acid monomers in a sequence alignment.
superfamily	A collection of protein families, thought to be related by *homology* (q.v.), but involving much more distant evolutionary relationships than those between members of a single family.
superfold	A very common protein *fold* (q.v.) shared by many protein *superfamilies* (q.v.).
SWISS-PROT	Database of confirmed protein sequences with extensive annotations. Maintained by the Swiss Bioinformatics Institute. Also see *TrEMBL*.
target	(1) The molecule recognized by a *probe* (q.v.). (2) The molecule recognized by a drug.
TCP/IP	Transmission Control Protocol/Internet Protocol. Protocols that control how data is packaged and reassembled (TCP) and addressed and routed (IP) over the *Internet* (q.v.).
threading	See *fold recognition*.
trace data	The raw data produced by automatic DNA sequencing.
transgenic	Containing additional (usually foreign or recombinant) genetic material in each cell.
TrEMBL	Translated EMBL. Database of protein sequences translated from the EMBL nucleotide sequence database. Not as extensively annotated as *SWISS-PROT* (q.v.).
UNIX	Very powerful multi-user *operating system* (q.v.).
unrooted tree	A phylogenetic tree in which the last common ancestor of all species on the tree is not shown.
UPGMA	Unweighted pair group method. A hierarchical clustering method used in phylogenetic analysis and the analysis of microarray data in which the distance between a merged cluster (*ij*) and any candidate data point (*k*) is taken as the simple average of (*ik*) and (*jk*). Cf. *WPGMA*.
upload	Transfer files from a local computer to a computer network or the *Internet* (q.v.).
URL	Uniform Resource Locator. A text address for a *web page* (q.v.). See also *IP address*.
UUE	UNIX–UNIX encoding. One of several methods for converting binary files into *ASCII* (q.v.). See also *MIME*.
VMS	Virtual Memory System, an *operating system* (q.v.) designed for older VAX computers and still used on some kCompaq (formerly DEC) file servers.

Web page (WWW page, W3 page)	A page of data written in *HTML* (q.v.), available over the *World Wide Web* (q.v.). See also *web site*.
web site (WWW site, W3 site)	A collection of organized *web pages* (q.v.) covering a common subject.
weight matrix	A statistical model used in sequence analysis in which each position in the sequence is modeled independently of all others.
Windows	Microsoft *operating system* (q.v.) with integrated low level operating system and GUI (q.v.).
World Wide Web (WWW, W3)	A way of exchanging data over the *Internet* (q.v.) using a *protocol* (q.v.) called *HTTP* (q.v.) and a program called a *browser* (q.v.).
WPGMA	Weighted pair group method. A hierarchical clustering method used in phylogenetic analysis and the analysis of microarray data in which the distance between a merged cluster (*ij*) and any candidate data point (*k*) is taken as the average of (*ik*) and (*jk*) weighted for cluster size. Cf. *UPGMA*.
XML	Extensible markup language. A *markup language* (q.v.) that not only specifies how text is presented but also allows the automatic description of embedded data objects.
X-ray crystallography	A technique for determining protein structures. Atomic co-ordinates are deduced from the way in which X-rays are scattered when they penetrate a precisely orientated protein crystal (cf. *NMR*).
Y-series	C-terminal fragments in a peptide ladder generated by *mass spectrometry* (q.v.).
Y2H	Yeast two-hybrid. A technique for detecting protein–protein interactions that is suitable for high-throughput analysis.
zip	Compressed *archive* format (q.v.) popular on PCs.

INDEX